Revolution in the Factory

Revolution in the Factory

The Birth of the
Soviet Textile Industry,
1917–1920

WILLIAM B. HUSBAND

New York   Oxford
OXFORD UNIVERSITY PRESS
1990

Oxford University Press

Oxford   New York   Toronto
Delhi   Bombay   Calcutta   Madras   Karachi
Petaling Jaya   Singapore   Hong Kong   Tokyo
Nairobi   Dar es Salaam   Cape Town
Melbourne   Auckland

and associated companies in
Berlin   Ibadan

Copyright © 1990 by Oxford University Press, Inc.

Published by Oxford University Press, Inc.,
200 Madison Avenue, New York, New York 10016

Oxford is a registered trademark of Oxford University Press

Library of Congress Cataloging-in-Publication Data
Husband, William.
Revolution in the factory: the birth of the Soviet
textile industry, 1917-1920 / William B. Husband.
p. cm.   Includes bibliographical references (p.
ISBN 0-19-506435-6
1. Textile industry—Soviet Union—History—20th century.
2. Soviet Union—Industries—History—20th century.   3. Industry and
state—Soviet Union—History—20th century.   4. Communism—Soviet
Union—History—20th century.   5. Soviet Union—Economic
policy—1917-1928.   I. Title.
HD9865.S652H87   1990
338.4'7677'0094709041—dc20   89-71028

1 2 3 4 5 6 7 8 9

Printed in the United States of America
on acid-free paper

With Love to Jeffrie

Acknowledgments

I am grateful for the opportunity to acknowledge many debts publicly. This is not the first work of Russian history to have its beginnings in one of Arno Mayer's graduate seminars on European history, and I value his unstinting encouragement and friendship. The late Vladimir Z. Drobizhev was exceptionally supportive of a project whose basic premises he did not fully accept and was always willing to share his unparalleled knowledge of early Soviet history. William G. Rosenberg has provided the proper mix of support and criticism at key stages of this project. Lynne Viola continues to be a valued critic, friend, and colleague. Richard S. Wortman, Moshe Lewin, and the late Cyril E. Black read this work as a dissertation and offered valuable insights. William J. Chase provided useful commentary on the parts of the manuscript he has read in various forms. Rifaat Abou-El-Haj, Kenneth K. Bailey, Toivo Raun, and Arnold Springer strongly influenced me during my early training.

A number of specialists and institutions have been instrumental in my research. I am but one of a generation of Princeton graduate students whose education and experience have been enriched by the patient tutelage of Orest Pelech, formerly of Firestone Library. I also thank the staffs of the Institute of Information on the Social Sciences of the Academy of Sciences of the USSR (INION), the Lenin Library, the State Historical Library in Moscow, and the libraries of the University of Illinois. I am indebted to the Soviet Main Archival Administration (GAU) for making available materials in Moscow and Leningrad vital to this study. In this regard I particularly thank L. E. Selivanova of GAU and Ivan Kurtov of the Institute of History. An International Research and Exchanges Board (IREX) fellowship and a Fulbright-Hays Fellowship For

Doctoral Research enabled me to spend 1981–1982 in the Soviet Union. I also thank the National Endowment For the Humanities for a travel grant in 1986 and Princeton University for fellowship support while a graduate student. I am grateful for permission to draw extensively from: "Local Industry in Upheaval: The Ivanovo–Kineshma Textile Strike of 1917," *Slavic Review* 47 (Fall 1988): 448–63; and "Workers' Control and Centralization in the Russian Revolution: The Textile Industry of the Central Industrial Region, 1917–1920," *The Carl Beck Papers in Russian and Eastern European Studies*, no. 403 (1985): 1–52. Oregon State University has facilitated the completion of the manuscript by providing various grants as well as a congenial and supportive atmosphere in its history department. I owe special gratitude to Nancy Lane as well as Stanley George, Susan Ecklund, and David Roll of the Oxford University Press for their consistently high level of professionalism. Rich and Edith Dallinger, Christine Lunardini, and Maureen Callahan have offered unfailing friendship and good cheer. I cannot begin to express adequately the contribution of my wife, Jeffrie.

I have employed the Library of Congress system of transliteration with slight modifications. Dates follow the calendar actually in use at the time. Dates thus lag thirteen days behind the western calendar before February 1, 1918 and conform with it thereafter.

Contents

Revolution in the Factory

I remember a group of workmen from a factory came to Lenine and asked how to run a factory. He held up his hands and said, "How do I know how to run it? I do not know how to run it. You go and try, and come back and tell me what you did, and then I'll try to learn from your blunders and mistakes and," he added humorously, "will write a book about it."

Albert Rhys Williams,
Testimony before the Senate Judiciary
Subcommittee on Bolshevik Propaganda,
February 29, 1919

1

The Russian Textile Industry between Revolution and Transformation

This is a local history of the Russian Revolution. The Bolsheviks seized national office in Russia in October 1917 with the vocal, albeit ephemeral, backing of the urban working class and the long-term objective of recasting the human condition in a more benign mold. To this end, they were armed with general guidelines for a transition to socialism,[1] a faith in imminent international proletarian revolution, and a reading of Marx that convinced them that their goals were scientific rather than utopian. Outside Russia's main urban centers and beyond the sway of the revolutionary intelligentsia, however, the consolidation of the revolution depended more heavily on immediate practical considerations than on ideology or slogans. Factory workers and local institutions persistently sought relief from the ongoing economic crisis that had already undermined the Imperial and Provisional governments. In outlying areas, many viewed the October Revolution as a license to redress long-standing grievances at the workplace on local initiative alone. Moreover, there everywhere existed support for taking revenge on the beneficiaries of the tsarist order and driving them from positions of privilege and responsibility, if not from the country itself.

The dynamics of revolutionary political and economic consolidation in 1917–1920, therefore, encompassed not only the aspirations of the Bolshevik revolutionary elite but also other significant pressures. The most important included the need to respond to the demands stridently being pursued by local constituencies; overcoming the resistance of the revolution's refractory enemies; finding and training, under crisis conditions, adequate staff for a new revolutionary administration; resolving conflicts among nascent revolutionary institutions; and reconciling pressing local needs with the long-term revolutionary goals being articulated by the national leadership. As a result, the local history of the early Soviet period is foremost an account of the centrifugal forces and fragmentation that the diverse expectations of the October Revolution brought unmistakably to the forefront. The present study of the textile industry of the Central Industrial Region[2] brings this dimension of the revolution into focus and, in the process, shows how the convergence of these conflicting pressures pushed the early Soviet system toward a more centralized mode of general operation.

Transforming the textile industry was a severe test of the Russian Revolution. When the Bolsheviks seized office in 1917 and during their consolidation of Soviet power in 1918–1920, the industry stood—like Russia itself—between a rural, peasant past and the prospect of a fully industrialized future. Formulating the mechanics of such a transformation had preoccupied all who advocated industrial modernization as the key to the nation's future—from government reformers to orthodox Marxists—since at least the 1890s. When tsarism fell during World War I, therefore, a broad range of socioeconomic prescriptions for the future, including those articulated at the grass-roots level, competed for primacy. Such ideological competition was, of course, inextricably entwined with the fundamental struggle for political authority that followed tsarism's demise.

It was in this arena that the Bolsheviks pursued the consolidation of political power and undertook their experiment in building a proletarian state and command economy. After the October Revolution, however, they faced an unbroken deterioration of the national economy, the broad cooperation on which they had counted failed to materialize, and they were nearly overwhelmed by a three-year civil war against a broad spectrum of enemies. Creating a new Soviet system, therefore, was a process punctuated by unanticipated problems and, of necessity, characterized by considerable improvisation. The Bolsheviks responded above all with a general concentration of authority in central institutions. The textile industry occupied a pivotal position in this process of adaptation, and the present work argues that class frictions, local and geographic loyalties, and collateral identifications such as family, gender, and self-preservation all combined to dominate the dynamics of revolutionary consolidation at the local level. The divergence of national and local aspirations in these areas laid the foundation for the friction between national and lower officialdom that became a hallmark of the Soviet state.

Historiography and the Textile Workers

Recent studies of the revolution and initial months of Soviet rule have clarified much of the complexity of the early stages of Russia's revolutionary transition. Revisionist histories of 1917 and the first half of 1918, which place strong emphasis on working-class experiences in society and at the workplace, have persuasively demonstrated the existence both of a more extensive political and socioeconomic consciousness among the population of the urban centers of the revolution by October 1917 and of broader popular support for Bolshevism than was previously acknowledged. Social historians have been particularly successful in documenting the role of a politically articulate "vanguard" of skilled workers in Petrograd and Moscow, especially skilled metal and railroad workers, in generating and mobilizing activism. In the process, these studies have transformed our perception of the Russian working class from that of a passive, malleable mass simply reacting to events to a force directly involved in the identification and defense of its interests and capable of producing its own frequently efficacious initiatives.[3] This interpretation has seriously undermined the primacy of intelligentsia politics in understanding the Russian Revolution as well as interpretations that rely on political manipulation or historical accident as explanations of the Bolshevik victory.[4]

Social historians, however, have yet to study thoroughly the local dimension of the revolution beyond mid-1918. Rather, in the process of substantiating their case that genuine working-class support legitimized the Bolshevik victory of October 1917, the revisionists of the revolution and first eight months of Soviet rule have been careful to separate the Bolshevik rise from the subsequent consolidation of power. While a class solidarity representing actual social

aspirations underlay the former, additional factors such as occupa-
tional, factory, ethnic, or regional loyalties—in addition to the
deepening of the national economic crisis—assumed the dominant
place during the latter. In the main, these revisionists suggest that
coping with the deterioration of class unity after October and
especially after mid-1918 contributed directly to the subsequent
excesses of single-party rule,[5] especially since the "vanguard" in the
factories in 1917 had been reassigned to other responsibilities.

There are, to be sure, Western studies that take a longer chrono-
logical view of the consolidation of Soviet power, but their chief
perspective on the post-October transformation is that of central,
national institutions.[6] These works on the party, state building, and
the body of policies identified as War Communism do not simply
neglect relations between the state or party and the general population and lower-level institutions. Most take into account, for example, the fact that before and after October support for increased
national authority often came "from below." In their quest for a
synthetic treatment and from their preoccupation with the themes
of centralization and ideology, however, they ultimately impose
upon their conclusions the concerns of the center. In the end, local
organs and constituencies speak only indirectly. Subordinates are
overwhelmingly evaluated not in terms of agendas they set for
themselves but according to their ability and propensity to fulfill
those perceived as premeditated by or imposed by circumstances on
superordinate institutions.[7] In addition, contributions by Soviet
émigrés preoccupied with finding evidence that political alternatives to Bolshevism existed after October 1917 have reinforced a
decidedly distinct but parallel propensity: to identify activity at
various levels in 1918–1920 directly with the politics of national
parties.[8]

Thus, recent works, although innovative in scope and resourceful
in marshaling documentation, fail to provide a sustained examination of any pivotal outlying area in the early Soviet period. This is a
serious gap in the historical literature. As already noted, much of
the revolutionary "vanguard," so pivotal in 1917, had dispersed by
mid-1918, thus changing the character of factory and low-level
politics fundamentally even in the capitals. Outside the main urban
centers, we lack a detailed study of any major area or industry,
making it impossible to gauge the degree to which we can apply our

knowledge of Moscow and Petrograd to the country as a whole. Hence, establishing the local perspective on 1918–1920 is important beyond its status as uncharted territory. Without a study of this type, we cannot fully assess the character of working-class politics beyond the sway of the proletarian "vanguard," nor can we understand fully the constituencies in whose name national institutions presumed to speak. The present work is a step in the direction of correcting this deficiency.

The Case of the Russian Textile Industry

This study maintains that the history of the textile industry from February 1917 through the end of 1920 represents a pivotal dimension of the revolutionary experience of Soviet Russia. The point is not that the story of the Central Industrial Region is "representative" because it conforms closely to the patterns of Petrograd and Moscow. Rather, the textile experience was "typical" of a large mass of Russian workers who, by definition, knew no other, and their numbers and consequence in the labor force were too significant to dismiss, in scholarship or in the politics of the era.

Indeed, the importance of the history of the textile industry of Soviet Russia is inescapable. Its scale of operations alone guaranteed that the industry would occupy a meaningful place in any effort to revive the national economy after World War I. Nationally, it was Russia's largest industrial employer in 1913, with the 690,124 workers in 1,449 textile enterprises accounting for 29 percent of the country's factory labor force.[9] Most of the industry was concentrated in the Central Industrial Region, which in 1913 contained in excess of 16 percent of the total number of Russian industrial enterprises and produced as much as 89 percent of gross output of the textile industry, or 37.1 percent of the nation's total gross industrial output.[10] The Factory Inspectorate counted 2,093,862 industrial workers in Russia by January 1, 1917, of whom more than 1 million were in the Central Industrial Region. Over 60 percent of these, largely women and adolescents, worked in textile factories.[11]

The centrality of the history of the textile industry in the revolution, however, does not lie in its scale alone. It was also, in a large sense, the country's most "Russian" major industry. Unlike the heavy industries, it was overwhelmingly financed by native capital.[12] In addition, its proprietors regularly extolled the peasant shrewdness that had enabled their ancestors to escape *muzhik* status in the not-too-distant past, and most were proud of their conservative business philosophy. Although owners flirted with state-of-the-art commercial practices, such as organizing syndicates and cartels, prevailing management habits resulted in the lowest level of vertical and horizontal combination of any major Russian industry on the eve of World War I.[13] The Bolsheviks would thus have to initiate much of their own momentum if they were to coordinate this important sphere of manufacturing on a national scale.

In addition, the composition of the labor force reflected the recent economic history of the country. Employing Tim McDaniel's categories of mass and conscious workers,[14] it would be difficult to find another major industry in which the distinction between conscious workers (with their higher degree of political consciousness and longer perspective on current events) and mass workers (more prone to respond to immediate circumstances and vulnerable to mood changes largely of circumstantial origin) was more meaningful. Unskilled and semiskilled workers,[15] a significant proportion of whom still maintained ties to the countryside, predominated, and women—traditionally perceived by both factory owners and revolutionary *intelligenty* as subservient at the workplace and in labor politics—constituted about two-thirds of the industry's labor force by 1917. Cohesion between the workers' representatives (largely revolutionary *intelligenty* and activists recruited from among the workers) and rank-and-file *tekstil'shchiki* was rudimentary and situational.

Beyond the striking attitudinal differences between conscious and mass workers, it is worth underscoring that the Russian factory labor force did not evolve from a well-defined artisanal tradition in the manner of the European working class, and the minimal training required for textile employment made this distinction still greater. The textile industry was not significantly represented in the large complement of artisanal workers found in Moscow at the turn of the century, and in this sense one can consider the textile work force even more "backward" than, for example, Russian metal-workers or construction workers.[16] Moreover, this was not an urban industry. Over 82 percent of the inhabitants of the Central Industrial Region lived outside cities and towns in 1917, and large-scale textile enterprises were similarly scattered, a factor that helped perpetuate the "peasant," that is rural, character of the industry.[17]

The experience of the textile industry in 1917–1921 also illustrates an important dimension of the political education of Russian local officials and mass workers. Since their industry encompassed large-scale factories as well as small workshops, employed a significant body of unskilled and semiskilled workers, and attracted larger than average numbers of those who retained their ties to the countryside, it could hardly have been otherwise. Unskilled and semi-skilled workers operated in every major industry, of course, but the paucity of skilled—and by extension conscious—workers in this industry gave it, in some cases by default, an independence of action and a uniqueness of orientation. In no way should this factor be underestimated. Others have noted differences between textile and other workers in the extent to which factory labor constituted a new socialization[18] and the radicalizing impact that propinquity with conscious workers from other industries could have on mass textile workers.[19] Diane Koenker, in particular, has carefully documented the greater political passivity of urban textile workers when physically isolated from more activist elements.[20] In the Central Industrial Region, we view mass workers left to a great degree to their own devices, with significantly less exposure to leadership and organization from conscious workers.

The results are illuminating. Unlike their urban comrades, mass textile workers in the provinces sometimes proved highly assertive politically. Although they could not create a sustained movement characterized as conscious by the standards of the time, their collective aspirations definitely shaped the tone of working-class politics in the outlying areas of the Central Industrial Region. The nonurban textile workers at times unhesitatingly adopted radical positions—transferring all power to the soviets, for example—that struck root in the cities more slowly. The critical point is not only that the textile industry encountered problems endemic to industry in more intense form. Rather, its prevailing characteristics made its

experience qualitatively different from heavy industries in urban
areas before and during 1917 and significant enough to require
special attention thereafter.

How the Bolsheviks dealt with these circumstances offers a direct
test of their capacity to convert rhetoric into action. In 1918–1920,
the Bolsheviks achieved comparative success in reorganizing this
industry, in part because the revolutionary government enjoyed a
rare opportunity to experience continuity of authority. The leadership of the Union of Textile Workers had traditionally supported
Bolshevism,[21] and about 85 percent of the textile enterprises were
located in areas continuously held by the Red Army during the
Russian civil war. This helped prevent the frequent reversals of
policy that plagued industries in which plants passed repeatedly
from Red to White control.[22] Moreover, workers occupied over 64
percent of the management positions in individual enterprises by
1920,[23] and they constituted over 50 percent of the members of the
directing collegia of Glav-Textile, the national textile regulatory
organ. Both figures were among the highest worker representations
in any industry.[24] As scarce textile products became vital both to
the Red Army and to a domestic market suffering a dangerously
low availability of manufactured goods,[25] therefore, efforts to re-
structure this industry revealed the full extent and limit of Bolshevik skills in artculating revolutionary priorities clearly and their
ability—and inability—to act on those priorities effectively. In
these circumstances, it should be noted, leading Bolsheviks such as
Ian Rudzutak, Alexei Rykov, and Lenin himself periodically
singled out the administration of the nationalized textile industry as
a model for other spheres of production.[26] Therefore, in the admit-
tedly difficult circumstances of the period, the case of the textile
industry provides a fair test of the efficacy of Bolshevik mass
politics.

Documentation

To study the Russian Revolution outside the cities requires documentation beyond the relatively rich materials for the urban areas,[27]
and this book introduces a broad range of archival evidence previously unexplored in Western historiography. It cannot have es-

caped the notice of the reader that the distinction between the Central Industrial Region and the city of Moscow cannot be complete, since the latter is obviously located within the former. Central and regional newspapers and journals, workers' memoirs, histories of individual factories, published documents, and especially Soviet secondary sources, although invaluable, are of uneven quality in drawing out the important differences, since they too frequently emphasize the perspective of Moscow. For that reason, this book draws heavily on records of local and central institutions of the textile industry located in the Central State Archive of the October Revolution (TsGAOR) and the Central State Archive of the National Economy (TsGANKh). Those examined systematically include directives from central agencies; communications between central and local organs; records of meetings and conferences at all administrative levels, including local gatherings; records of meetings and communications at the factory level; and assorted questionnaires, reports, and correspondence generated both locally and centrally. To a lesser degree, the study also draws on limited access to assorted relevant materials in the Central State Archive of the October Revolution and Building of Socialism in Leningrad (TsGAOR SSL) and the Central State Archive of the City of Moscow (TsGAgM).

Premises

This study holds, above all, that between February 1917 and the end of 1920 the workers and their representatives in the textile industry of Russia's Central Industrial Region displayed strong support as a group for the revolution *as they perceived it* but a much lower degree of class solidarity than is presently attributed to the workers' movements of Petrograd and Moscow, particularly prior to the October Revolution. While class unity coalesced in 1917 and then disintegrated after October in the capitals, it never significantly transcended localism in the outlying regions of the Central Industrial Region, where the unskilled and less politically conscious predominated. Hence, the textile workers' organizations strongly supported the Bolshevik interpretation of working-class aspirations in the second half of 1917, and in some cases textile workers in the periphery anticipated the radical demands of the more conscious and cautious workers' movements in Petrograd and Moscow. Local identifications, a preoccupation with economic concerns, and a definition of revolution that stressed the immediate redress of past grievances dominated their expectations of "revolution," however, throughout 1917–1920.

By extension, this work also argues that the experience of the textile industry encompasses in microcosm important political scenarios that made their way to the national political forefront by 1920–1921. To cite one central example, the use of "bourgeois specialists," that is, those with professional expertise who had served during tsarism, was an ongoing source of local discontent that national leaders found themselves continually having to explain and justify. In the textile industry, incessant pressure for the removal of "bourgeois specialists" resulted from the widespread expectation that the revolutionary transformation could be equated with an immediate reordering of influence and authority.[28] To call attention to a different instance, the tenacity with which institutions at all adminstrative levels laid claim to and then defended their jurisdictions foreshadowed similar battles at the highest reaches of the party at the end of the civil war. The struggles fought in the national arena in 1920–1921—Workers' Opposition, Democratic Centralism—had their roots in the process of local consolidation of the revolution.

In addition, the present work contends that the strength of administrative institutions in 1917–1920 and the importance of party affiliations need to be reassessed. The gap between the ideal functioning of institutions articulated in state and party directives and actual daily operations was wide. Behind grandiose directives issued in the name of this or that body, one finds organs that formed, reorganized, and dissolved with a regularity that could not but undermine minimal operation, not to mention efficiency. Governing boards and directorships often consisted of a single literate person or small group sometimes unwillingly drafted to responsibilities for which they had little inclination or preparation. Institutions that expressed themselves assertively and authoritatively one day might not exist the next. Moreover, evidence of the importance of party affiliations is all but absent from local records. Local constituencies, although supportive of the revolution itself, articulated their aspirations in terms of specific needs, not according to the political identifications and programs that divided the revolutionary intelligentsia.

Finally, this study maintains that the frictions between national and local officials, which became critical at the end of the 1920s, were an important aspect of the Soviet system from the time of the October Revolution. New Western works on the 1920s, the period of the First Five-Year Plan, and Stalinism in the 1930s—although at odds in important ways—share a common and pronounced concern with the dynamics among central authorities, local officials, and the rank and file. This focus has, at the very least, added an important dimension to our understanding of early Soviet history and, in its most controversial forms, challenged conventional wisdom about the fundamental character of the Soviet system itself.[29] A close examination of the Central Industrial Region in 1917–1920 reveals pronounced evidence of mutual antagonism between national and subordinate authority from the outset. This important relationship in Soviet politics emerged not during the power struggles of the period of the New Economic Policy but was rooted in the very character of the Bolshevik seizure of power.

2

The Legacy of Tsarism
and Revolution

The textile industry of the Central Industrial Region consistently frustrated Russian reformers long before the Bolsheviks attempted its reconstruction. Complexities were rooted in the chronically poor conditions of life and work—even by the generally low standards of tsarist Russia—that had long exacerbated the raising of labor productivity, increasing efficiency, and improving the quality of life of the labor force. Conscious *and* mass workers in the industry, however, were far from simply victims of their past or of socioeconomic determinism. In 1917, even mass workers proved capable of identifying their interests, conceiving defensive strategies, influencing the behavior of the conscious activists among them, and at times formulating initiatives. Indeed, Russian textile workers pursued an agenda that far more reflected material need and local concerns than the programs and party identifications that preoccupied the revolutionary elite. In these conditions, the collateral loyalties of individual party members even undermined intra-Bolshevik cooperation, which could in no way be taken for granted.

Prerevolutionary Conditions in the Russian
Textile Industry

The manufacture of textile goods occupied an important place in the economy of prerevolutionary Russia. As early as 1051, the Code of Iaroslav prescribed a penalty for stealing petticoats, and in the sixteenth century the marketing of textile products began on a small scale.[1] At the beginning of the eighteenth century, the military projects of Peter the Great significantly expanded the demand for textiles, especially woolens. Cottage craftsmen principally met these requirements, although several nonmechanized manufactories also appeared in Moscow by the end of Peter's reign. Cotton spinning and weaving expanded as a broadly based *kustar'* (handicraft) industry in central Russia during the Petrine period as well. By the middle of the eighteenth century, therefore, cloth manufactories generally utilizing twenty to thirty looms were established throughout Moscow and Vladimir provinces to process the semimanufactured goods produced by cottage workers, and by the end of the century more new manufactories were being founded outside the town of Moscow than within it.[2] Those who toiled in these workshops were largely state and manorial peasants or *otkhodniki*, serfs who departed their villages seasonally with the permission of their lord, often traveled and contracted wages as a collective *artel'*, returned home regularly, and used their earnings principally to discharge *obrok* (quit-rent) obligations. Both concentrated and *kustar'* industry continued to coexist beyond the nineteenth century.

The mechanization of cotton spinning in the 1840s, however, began to shift the balance definitively in favor of large-scale factory production. Once England began to allow the export of textile machinery in 1842, Russian enterprises increased in size, capital concentration, and importance as sources of employment. In this process, cotton displaced wool as the most important textile product. In addition, a "gathering" of workers in the mills and a concomitant sharp decline in the proportion of output of all textile products from the *kustar'* sector followed the emancipation of the serfs in 1861.[3] By 1914, 89.1 percent of those involved in cotton production worked in factories employing more than 500 laborers, and only 1.6 percent worked in shops of fewer than 100. The figures for other spheres of textile manufacturing were only slightly less polarized.[4] By the end of the nineteenth century, large-scale mechanized enterprises concentrated in Moscow, Vladimir, Kostroma, Tver, and Kaluga provinces and owned by a relatively few families dominated the industry, and only England, the United States, and Germany surpassed Russia in textile output.[5]

This transformation of the industry in the nineteenth century created a hybrid hired labor force of peasant-workers. On the one hand, the maintenance of legal distinctions increasingly at odds with reality caused government statistics on the work force to list as peasants many who had but a passing acquaintance with life in the countryside. In actual fact, over 90 percent of the textile workers were committed to full-time factory employment by the final decade of the century.[6] By the early years of the twentieth century, over half the labor force in key Moscow mills consisted of second- and even third-generation workers. On the other hand, the "peasant" character of the Russian textile worker was considerably more than a statistical anachronism. The fact that the majority of textile enterprises were located outside of the major cities limited the cultural transformations associated with urbanization.[7] In the urban mills, the propensities of many inhabitants of factory barracks, in conjunction with the owners' use of guards to control ingress and egress, curtailed contact with the influences of city life. The semiskilled and unskilled, who generally were more prone to maintain their rural ties,[8] predominated in the industry, and their high degree of common experience in geographic origin, language, and religion kept both urban and rural textile workers largely within the peasant subculture. Large numbers of textile workers visited the countryside during the Easter and summer shutdowns of the mills, fled there in times of adversity, and in some cases retired in the village. The *artel'* hiring pattern was still in evidence at the end of the nineteenth century, and *zemliachestvo*, another practice rooted in peasant life, survived the Bolshevik Revolution. The *zemliachestvo* tradition, which linked openings in specific trades, shops, and enterprises to particular villages, facilitated hiring and eased the transition from rural to factory life but, by definition, reinforced the village tie even after arrival in town.

Women came to preponderate numerically in the industry in the final quarter of the nineteenth century, only partially as the result of mechanization. Women had constituted less than one-fifth of the textile work force at the end of the eighteenth century,[9] but by 1913 they held more than half the jobs in the industry. Mechanization had eliminated physical strength as a prime professional consideration and made it possible for females to operate spindles and looms as productively as males, but this alone did not convince mill owners to hire them in large numbers. More significant was the fact that women brought from their peasant background a legacy of subservience and obedience, enabling management to pay them lower wages and to perceive them as less threatening politically.[10] Indeed, factory inspectors noted the particular determination of textile proprietors to "replace dismissed workers only with women" in the years following the Revolution of 1905.[11]

While a docile work force may have been a general desideratum among industrialists beyond the borders of Russia, the conjunction of large numbers of female peasant-workers, a highly paternalistic management philosophy among the owners, and the particular requirements of textile production gave rise to especially restrictive factory conditions in the Central Industrial Region. A government-sanctioned system of fines, for example, provided important leverage for all Russian factory owners both to impose discipline and to reclaim a portion of wages, but the textile industry magnified the situation. The enforced absences due to the responsibilities of motherhood for such a large number of female workers created additional excuses to levy fines, while the very time spent away from the job reduced the piece-rate wages paid. In addition, all Russian workers could be searched when leaving their factory, but the exaggerated conscientiousness with which many managers and foremen discharged this task toward women attached additional humiliation to an indignity already deeply resented. Moreover, fragmentation existed among the textile workers themselves. Patterns of advancement, to cite one factor, did not conform to those in other industries. Skilled male workers prevented women in textile manufacturing from passing regularly through the apprentice–journeyman–master curriculum, and females typically worked in semiskilled tasks at best for the duration of their working life. In addition, after the beginning of the twentieth century male activists systematically excluded the female majority from labor politics, and male textile workers habitually treated their female counterparts with hostility on the job. Gender thus combined with the barriers erected by differing levels of skill, shop loyalty, and so on as divisive factors within the labor force.

There were also, nevertheless, expressions of common interests. Male and female textile workers alike regularly complained about managerial disrespect. They resented familiar forms of address, the refusal of many foreign supervisors to learn the Russian language, and the capriciousness of the system of lines. During textile strikes, workers' demands commonly included a declaration against sexual harassment by foremen and directors. The fact that responsible personnel regularly withheld the few available services provided in some factories, such as medical care, penalized all workers, and supervisors even found ways to exploit village ties, as by assigning the best machines to those who brought eggs and other gifts from the countryside following holidays.[12]

In addition to oppressive labor conditions, the general circumstances of factory life adversely affected the textile work force. Low levels of education, poor food and quarters, substandard sanitation and health facilities, and general fatigue undermined spirit, dignity, and morale. Long hours were ubiquitous. In the 1870s, wool workers labored from four o'clock in the morning until nine at night. When labor legislation in 1885 legally barred women and children from night work, owners scheduled a fourteen-hour day shift. Although by 1913 the average workday for adults throughout the industry purportedly averaged ten hours, women in the massive Moscow Prokhorov Trekhgornaia Mill typically left their barracks at half past three in the morning and returned at ten at night.[13] Living areas were crowded and dirty. Owners of large urban mills quickly recognized the advantages of providing barracks and food themselves. Sleeping in one's shop, on a desk, or in a crowded and poorly ventilated dormitory could consume as much as one-third of the pay of an unmarried worker. Two married couples with children commonly shared quarters that measured two and one-half by four meters. Such units contained but two beds, and as many as eight children slept on a trunk or the floor. At the Trekhgornaia, meat appeared only at Christmas and Easter, and staples of the diet consisted of "*kasha* with *shchi, shchi* with *kasha*." As many as eight people ate from a common dish, and six drank from the same cup.[14]

Illness and ignorance also seemed self-perpetuating. In work areas, the dust emitted by spindles in unventilated shops took a serious toll on health, and all but a few were physically exhausted by age forty. The absence of proper safeguards around equipment caused the Factory Inspectorate regularly to cite the textile industry as a leader in industrial accidents. Literacy, an important component of professional skill and social consciousness, slumped well beneath levels in other industries.[15] In these circumstances, alcohol abuse proliferated despite stiff fines for being caught. While women might seek refuge from the drabness of factory life in religion, the nearby tavern provided the main male diversion, and those that extended credit were especially popular. The Easter and Christmas breaks typically were drunken sprees.[16]

Such conditions fostered discontent but did little to raise political consciousness or stimulate concerted action. Western and Soviet scholars have extensively documented the textile workers' conspicuous participation in strikes, many of which were large, well coordinated, and highly disciplined.[17] Evidence of political volatility abounds: the textile workers' high profile in the major strikes of the late nineteenth century and in the Revolution of 1905; the rise in the number of strikes, strike participants, and lost workdays in the industry in 1908–1913;[18] labor discontent during the war; and the strong visibility of textile disturbances in February 1917.[19] Textile workers, nevertheless, were poor candidates for sustained action. Those who lived in urban factory barracks were difficult targets for outside agitation, and early in the twentieth century they created fewer disturbances than their nonurban counterparts. Rural textile workers, on the other hand, tended to be more activist, but geographic dispersal made it difficult to coordinate their actions.[20] Both groups focused more strongly on immediate grievances than on long-term considerations. Thus, government repression effectively curtailed the nascent union movement in the industry after 1905, and in spite of renewed activism in 1908–1909 poor communication and low political consciousness prevented the regeneration of industry-wide workers' organizations. At the time of the February Revolution, the Union of Textile Workers counted but 64 active members.[21] There were those, to be sure, who had continued to gather political experience through unbroken commitment, despite the demise of opportunities for organized resistance after 1905.[22] A significant share, however, even of those who would hold positions on factory committees in the industry lacked pre-1917 political experience and were inactive without outside motivation.[23] Indeed, the definition of an activist in 1917 and thereafter is quite elastic, including the small number of those with prerevolutionary political experience and a broad spectrum of those politically inactive before the fall of tsarism. Local accounts communicate that many who held positions of responsibility after the revolution did so for reasons unrelated to politics or political consciousness (such as that they were the only available literate person).

There also existed a low level of coordination among ownership that evoked comparisons with industries at an earlier stage of development, despite the scale of textile production by World War I and its high degree of capital concentration. An inner circle of native entrepreneurial families based largely in Moscow—and proud of their differences from the foreign-influenced manufacturing and commerce of Saint Petersburg—dominated the proprietorship of large-scale enterprises. In the nineteenth century, a high consciousness of common economic interests, social legacy and status, and cultural and Old Believer religious outlook had caused these industrialists to develop a corporate identity that they nevertheless failed to mold into a consistent political instrument. Textile magnates largely eschewed politics not directly tied to business affairs and successfully resisted state regulation of their industry, but at the same time expected government assistance in times of crisis. This pattern began to change only slowly in the final decades of tsarist rule. Scions of the leading textile families led a proportionally small but vocal cohort of Moscow industrialists whose participation in public life culminated in assertive participation in Duma politics, War Industries Committees, and national and regional industrialists' organizations. Although often supportive of the liberal reforms of the period, they were motivated as much by the realization that current practices and conditions inhibited growth and productivity as by ideology or humanitarianism. To these ends, part of their effort was to replace the personalized approach to ownership and management, whereby the will of the owner was the guiding principle in each factory,[24] with economically rational, collective action.

These reformers never molded a consensus in the industry, however, and the majority of groups of large-scale enterprises continued to be run as family proprietorships. This precluded the coordination of production and distribution characteristic of a similar level of capital concentration in other Russian industries and in the West. It is true that in the years prior to World War I major producers periodically attempted to control prices and distribution through syndicates and cartels, but such projects failed to affect the rhythm of the industry in the same manner as in, for example, Russian mining and metallurgy. This was above all due to the high propensity of textile industrialists to abstain from or abandon collective efforts in favor of independent action as it suited them.[25] Moreover, the linking of industry and investment followed a special pattern in Russian textile manufacturing. In western Europe, bankers as a matter of course took positions on the directorships of enterprises in which their institutions invested heavily. In the Moscow financial establishment, by contrast, textile magnates owned banks or held controlling shares and often served on bank directorates, thereby surrendering no flexibility.[26] It is no exaggeration to say that family alliances and geographic propinquity were more important than capital concentration in the owners' organization of production.

World War I caused significant displacement in the industry without fundamentally altering its character. As a group, factory owners differed little from before the war: adept at turning a profit but inept or indifferent when facing long-term considerations. After an initial disruption in 1914, state military orders produced a 61.9 percent increase in profits in the next two years.[27] Supporters as well as critics of the tsarist state continually charged that collusion on pricing and distribution existed between the magnates and the government, and the weak attempts by the state to regulate the industry during the war did nothing to dispel these perceptions. In point of fact, government regulation did enlarge profits. In 1915, the Ministry of Trade and Industry created Centro-Cotton (*Tsentrokhlopok*) and Centro-Wool (*Tsentrosherst*), in which manufacturers participated, to regulate supply and production. These committees fixed prices of materials, such as raw cotton, but not of semimanufactured goods, such as cotton thread and yarn, and finished goods. Hence, spinning mills purchased raw materials at controlled rates, but set their own prices for output. Consequently, speculation and profiteering occurred at this and every subsequent stage of the production process, and the increases were eventually reflected in the cost of the scarce finished product. Similar regulatory failures occurred in wool and flax processing, and in the end government agencies exercised but scant authority over the distribution of goods.[28]

The war also reinforced the processes that directly lowered the workers' morale. The proportion of women in the labor force grew from one-half to two-thirds by 1917,[29] and during the war a small minority entered skilled work almost completely closed to females previously. Affecting a larger number, however, was the sharp rise in the number of unattached women who now relied entirely on their meager textile wages for survival. To complicate matters, real wages in the industry fell to a fraction of their prewar level.[30]

Consequently, strike activity erupted again in 1915–1916, but the textile workers retained their disjointed prewar political character. In May 1915, the textile centers of Ivanovo–Voznesensk and Shuia (Vladimir province) experienced a general strike. When it spread to Kostroma in June, the police fired on demonstrating workers. When the weavers of Ivanovo–Voznesensk quit work again to hold a rally on August 10, tsarist authorities once again shot demonstrators.[31] Nevertheless, memoirs, even of those who later became activists, consistently deny to the war experience a significant role in creating a more unified consciousness or impulse toward political participation.[32] In 1917, textile workers in outlying areas continued to rely on their established pattern of responding primarily to local grievances. In the city of Moscow, the textile worker would actually display a lower propensity to strike, be more prone to demand higher wages than control of the factory or political concessions, and even pass fewer resolutions than would workers in other major industries.[33]

The Year 1917 in the Central Industrial Region

By February 1917, therefore, the textile industry presented an impediment to the revitalization of Russia that any postwar government would have to overcome.[34] At the time of the overthrow of tsarist authority, however, the *tekstil'shchiki* did not yet function as an independent and unified force and shared the political stage with activists from other industries and representatives of other classes. In the Central Industrial Region in February–March 1917, workers from other industries—the 150,000 employed in metal fabricating and machine construction as well as railroad workers—also played a leading role, and in every case the winning of the allegiance of the soldiers was critical. The propertied classes, meanwhile, generally viewed themselves as the legitimate inheritors of power and organized their forces to pursue it. It is true that the major crises in

Petrograd did generate direct responses in the Central Industrial Region, as demonstrations engulfed the region in April, June, and July. On the other hand, regional politics did not simply replicate those of the capital. Thus, despite some common elements—mass strikes and demonstrations, the institution of dual power, the arrest of tsarist officials, the lack of authoritative centers furnishing revolutionary direction, a paucity of declarations on the war—there was no typical transformation in the region either in February or thereafter. Local variations predominated.

The experience of the city of Moscow was significantly different in February from that of Petrograd. In Moscow, a relatively low concentration of industrial laborers reduced the political impact of the factory workers as artisans, service workers, white-collar employees (*sluzhashchie*), and organized entrepreneurs strongly influenced the rhythm of developments. Thus, when Moscow observed International Women's Day on February 23 with meetings and public gatherings, it was not with the sense of immediacy evident in Petrograd. Radicals disseminated revolutionary slogans, but at the time local conventional wisdom anticipated no significant change before the end of the war. Support for the revolution, nevertheless, grew rapidly under the influence of developments in Petrograd. On February 25, as the disturbances in the capital reached their critical stage, government authorities felt it necessary to break communication lines between the two cities in an attempt to contain the unrest. Nevertheless, by February 27 word of the Petrograd developments spread throughout Moscow, and on February 28 simultaneous demonstrations riddled different sections of the city despite the absence of any crystallized, coordinated leadership. The Moscow garrison came over to the revolution on March 1, and on the following day crowds stormed the jails, released prisoners, and arrested Governor N. Tatishchev and other leading officials.

Dual power took shape as in Petrograd, but with a closer working relationship between citywide institutions initially. First, the Committee of Public Organizations (*Komitet obshchestvennykh organizatsii*), originally 150 representatives of all elements of the Muscovite population, formed on the evening of February 27 as the analogue of the Provisional Government, and organizations of all political persuasions formed corresponding committees in each of the eleven *raiony* of the city. Second, elections to the Moscow

Soviet of Workers, although far from uniform in proportional
representation, were held on February 28 and March 1, when the
Soviet actually began work. Direct elections in the factories, or in
some cases indirect elections by factory committees, led to the
immediate formation of *raion* soviets throughout the city as well. In
addition, organs representing class and professional interests took
shape as battle lines whose full significance was not yet appreciated
formed. The textile magnate P. P. Riabushinskii chaired the Organizational Committee of the All-Russian Union of Trade and
Industry on February 27, while on March 15 17 representatives of
various unions—metalworkers, printers, woodworkers, construction workers—met to form a Central Bureau of Trade Unions. The
Soviet established a special section to lead the organization of lower
soviets, factory committees, and trade unions, but the step proved
unnecessary. Local initiative had already put this into motion, and
the formation of factory committees in particular, already under
way before February, accelerated dramatically and well in advance
of the revival of trade unions. In another significant step, the
Moscow Soviet, over the heated objections of *raion* representatives,
issued a ban on strikes that won wide compliance by the middle of
March, although workers continued to press their economic demands and widely implemented the eight-hour workday on their
own authority.

Outside the city, local developments followed the outline of this
basic pattern, but by no means did outlying areas simply emulate
the "second capital." In Nizhnii–Novgorod, a walkout of 23,000
workers of the Sormovo railway, steel, and mechanical plants on
March 1 precipitated the formation of an "all-class" Executive
Committee, while workers' representatives also created a temporary
Soviet of Workers' Deputies. Neither institution could actually
control events, however, even after the garrison defected and the
provincial governor and other significant officials were arrested. In
Tula, where the 30,000-member garrison had helped repress the
local strike movement at the beginning of 1917, metalworkers led
more than 40,000 workers from two arms and cartridge factories in
a demonstration on March 2, with key support from railway
workers. By March 4, an Executive Committee representing organizations from all classes and the Soviet of Workers' Deputies led by
Mensheviks and Socialist-Revolutionaries began operations, both

committed in the short-term to the reinstitution of order and the recommencement of work.

Proceedings engendered equal or greater volatility in the textile centers. In Tver, the Morozov textile factory served as a meeting point once a strike began on March 1, and activists carried news to rural enterprises and military barracks. On the following day, crowds of workers and soldiers killed Governor von Biunting and General Chelovskii, and the commander of the cossacks took his own life. On March 3, the Soviet of Workers' Deputies, which coexisted with an "above-class" Executive Committee, began to function with a declaration of support for the convocation of a Constituent Assembly. Early cooperation between the two organs reflected a mood more intransigent than in many other localities, as in their unified rejection of the Provisional Government's order to release arrested tsarist authorities.

By far the most important political center of the region after Moscow was Ivanovo-Voznesensk, the "Russian Manchester" that prevailed economically over the Ivanovo-Kineshma region.[35] The city of Ivanovo-Voznesensk was important not only as an industrial center but also for its political symbolism. The site of Russia's first soviet in 1905, the city was a revolutionary bastion of long standing and also a significant center of prerevolutionary Bolshevik support. News of the Petrograd events did not reach the city until the evening of March 1, but by the following morning activists from the large Kuvaev textile plant already organized a public demonstration in the square opposite the City Duma. In the afternoon, the crowd marched on the Neburchilov Barracks, in which the military commander Smirnov, one of those responsible for firing on demonstrating workers in 1915, had locked the 199th Infantry Reserve Regiment. The soldiers joined the population in the streets the next day. When the Ivanovo-Voznesensk Soviet opened on March 3, the Bolshevik V. P. Kutuzov assumed its chairmanship. The mood of dual power, however, prevailed here as well. The Soviet possessed the strength to seize power from the outset, especially in view of the extreme hostility between workers and the propertied classes in the city, but exercised caution. Its chief slogans—Constituent Assembly, eight-hour workday, a "democratic republic"—echoed those issued elsewhere, and the Soviet agreed to share authority with representatives of other public organizations, including the City

Duma and industrialists' groups. This led to the formation on the same day of a Committee of Public Safety, which placed the return to normalcy at the top of its agenda. None of this prevented the arrest of local police officials, but in the main the Soviet largely occupied itself after March 4 with trying to effect the resumption of work in the mills.

A similar lack of uniform experience appeared in the smaller Ivanovo–Kineshma towns. As in Ivanovo–Voznesensk, textile workers played a key role in the opening of the local soviet in Kostroma and in pressuring the City Duma to share power in the Kostroma Joint Committee of Public Safety. In the city of Vladimir, where commerce rather than industry dominated the economy, the Provisional Executive Committee allotted the workers' representatives the role of junior partners. In Kovrov (Vladimir province), demonstrations instigated by skilled railroad workers and the 299th Infantry Reserve Regiment led to the formation of the local Soviet, which shared authority with a Provisional Citizens' Committee. In Vichuga (Kostroma province), on the other hand, the 35,000 workers of eight closely clustered textile factories held mass demonstrations and opened a Soviet by March 4, which, despite its Bolshevik majority, declared itself a nonparty institution protecting the political and economic interests of all workers. In Kineshma (Kostroma province), another manufacturing center, Menshevik activists assumed the leading roles. On March 2, 15,000 people elected an "all-class" Revolutionary Committee of Public Safety, and a Soviet began work the following day. In Nerekhta (Kostroma province), a dual-power Provisional Executive Popular–Revolutionary Committee began operations on March 5. Almost impossible to chronicle in detail are the remote areas, some of which contained only a single large-scale textile plant, which often lacked newspapers or regular lines of communication. There owners and officials typically withheld the news of the revolution as long as possible, with mixed results.[36] As in the more settled areas, an uneasy truce evolved, but friction within the enterprises generally increased after the initial conciliatory mood of the February Revolution subsided.

This lack of a standard response to the February events set the tone for the remainder of the year. In some cases, of course, developments in the region were in full accord with concerns prominent in Petrograd or Moscow. As elsewhere, the institution of the eight-hour workday dominated short-term considerations in both textile and metal fabricating centers during the spring and gave the nascent workers' movement its initial orientation and cohesion. The slogans carried in the May Day demonstrations in Nizhnii-Novgorod and Ivanovo–Voznesensk on April 18 mirrored the concerns of the capitals and served notice that the tolerance of February was giving way to a concentration on interests of specific importance to the workers. In a separate incident, a Soviet-sponsored demonstration in Petrograd on June 18 became dominated by Bolshevik slogans, and the same thing occurred in Kovrov and Nizhnii-Novgorod. Following the Kornilov affair, a muddled attempt to use the army to restore order in Petrograd in August, a Bolshevization of the soviets throughout the provinces of the Central Industrial Region paralleled the same phenomenon in the capitals.

In other cases, however, developments in the capitals largely bypassed the periphery. On March 27, for example, the Provisional Government issued a "Declaration of War Aims," which, although ambiguous, denounced annexations and foreign domination as goals in the war. When a note of explanation to the Allied governments by Kadet Minister of War Paul Miliukov—that undercut the declaration—became public in mid-April, several days of demonstrations on both sides of the war issue erupted. Responses to the Miliukov note were not widespread outside the capitals, however, although the crowd in the textile center of Tver was estimated at 75,000.

One distinctive characteristic of the textile-producing areas, however, was that working-class organizations outside Petrograd and Moscow rapidly became more assertive than the corresponding institutions in the capitals, and in key instances they anticipated urban developments. "All power in Orekhovo–Zuevo [Moscow province] is in the hands of the workers. The Petrograd Soviet lags behind and fails to defend the interests of the workers and soldiers," declared the delegate Efimov to the Seventh Bolshevik Party Conference in Petrograd in April.[37] If anything, the rioting over the issue of the Soviet seizing power in Petrograd, known as the July Days (July 3–5, 1917), produced even more strident postures in the Central Industrial Region. The Petrograd and Moscow Soviets resisted overtures from the crowds to seize authority, but this did not reflect the mood of outlying areas. In Ivanovo–Voznesensk, where 25,000 to 30,000 demonstrated on July 3–5, the local soviet passed a resolution favoring the transfer of all power to the soviets, in contrast to developments in both Petrograd and Moscow. The Ivanovo–Voznesensk Bolshevik organization later reported to the Sixth Party Congress in August that following the July Days the local bourgeoisie and intelligentsia had been unsuccessful in attempts to repress the party after July, since the workers continued to rally to the Bolsheviks.[38] From Orekhovo–Zuevo, where another 25,000 demonstrated on July 3–5, the mood was reported as sympathetic to the local soviets but antagonistic to those of Petrograd and Moscow,[39] and a meeting following the general demonstration endorsed the transfer of power to the soviets as well. An additional 15,000 turned out for July Days demonstrations in nearby Shuia, and further demonstrations continued in Ivanovo–Voznesensk and Shuia through July 6–7. The Bolshevik leadership of Moscow *Oblast'* believed that the Provisional Government was less able to use force in the cities of the Central Industrial Region after July, a factor that increased opportunities for militancy there. Finally, as the news of the October Revolution spread: "In Ivanovo–Voznesensk itself there was no talk of a transfer of power since in reality it was [already] in the hands of the Soviet of Workers' and Soldiers' Deputies, the city council [*uprava*] and the union of textile workers."[40] While the full meaning of such information cannot be assessed unskeptically, the greater power of the soviets outside the capitals by the eve of the October Revolution was generally conceded. Moreover, significant impetus for their militancy came "from below."

Between July and October, the textile workers' movement outside the city became more strident. Moscow workers' organizations became more disciplined throughout the year, but in the process they fell out of step with the greater militancy of the outlying textile centers. The city's ability to attract a critical concentration of experienced veterans of the prerevolutionary underground, as reflected in the formation of the Central Bureau of Trade Unions in Moscow on March 15, in due course tempered its sense of revolutionary adventurism. The Central Bureau of Trade Unions did launch a general strike on August 12 in protest of the opening of the Moscow State Conference, but this broadly discussed and even equivocal protest was hardly a manifestation of unbridled militancy. By August 29–30, the leaders of forty-seven unions in Moscow were working together on Bureau of Trade Unions' projects such as the formation of Red Guards, and on the eve of October the city labor movement effectively curtailed radicalism in favor of the long-term considerations on the horizon. It would, of course, be a gross mistake to equate the actions of these union leaders with the mood in the Moscow factories, especially in the textile industry. There is no question, however, that there existed in Moscow a greater potential to direct organized labor activity away from the more immediate priorities of the mass worker, especially in light of the textile workers' relative passivity toward organizations and resolutions there. As we shall see more completely in the following section, mass textile workers outside the city continued even after the October Revolution to pressure their representatives to base their actions around the twin goals dominant in their factories: economic betterment and redressing long-standing grievances. Moreover, the textile workers' movement crystallized in the area where textile manufacturing prevailed (largely Ivanovo–Kineshma plus the eastern portion of Moscow province), leaving metal-workers and others to work independently in their own strongholds such as Tula and Nizhnii–Novgorod.[41]

This was the backdrop against which the Bolsheviks began to transform themselves from an underground to a mass party, and despite the existence of conspicuous Bolshevik support in the Central Industrial Region the party was in no position simply to command a monolithic following. As numerous studies have established, the Bolsheviks did not possess an extensive and unified national apparatus, articulate a detailed program of revolutionary transformation, or dominate socialist politics before October 1917. Party consciousness itself was slow to develop among the working class in 1917 even in the cities.[42] It would not be unfair to say, moreover, that at the time of the unexpected February Revolution the common elements in the various Bolshevik positions on the future of Russia combined a particularly deterministic interpretation of Marx's view of historical development with general objectives from the program of the Second Party Congress of 1903, many of which were common to the underground labor movement as a whole.[43] This changed but gradually following the overthrow of tsarism, since the initial propensity among socialists to work as a coalition prevented the leading Bolsheviks in Petrograd, and in the Central Industrial Region as well, from articulating a specific and independent program. When Lenin began to call for the formulation of a new party program after the February Revolution,[44] therefore, his assessment of the present situation moved him to propose gradual transitional guidelines rather than the rapid implementation of socialism according to well defined steps. His *April Theses*, while advocating accelerating the bourgeois February Revolution into a socialist one, stressed that "it is not our *immediate* task to introduce socialism, but only to bring social production and the distribution of products under the *control* [*k kontroliu*] of the Soviets of Workers' Deputies."[45] Indeed, placing existing capitalist institutions under the *kontrol'* (i.e., accounting and supervision) of the working class would dominate his socioeconomic pronouncements in the first half of 1917,[46] even though the party would not actually add "workers' control" to its slogans until May.[47]

If a socialist revolution were possible in the near future, however, the Bolsheviks would need both to unify their position on *kontrol'* and to define concretely the authority of the institutions that would implement it after the revolution. This was no small task, and one in which the party would have to adapt itself to mass developments. Indeed, factory committees had already begun to define workers' control for themselves based on experiences in economic warfare with the factory owners. In large measure, this entailed taking the steps necessary to keep factories in operation. In extreme cases, it could also result in transferring full authority to the factory committee itself, a position close to the anarcho-syndicalist view that workers should directly confiscate enterprises. On the other hand, the Mensheviks—in keeping with their view that further capitalist development would be needed before a socialist revolution—championed "state control," with existing capitalists playing the leading role in the economy for a protracted but unspecified period of time. Among the Bolsheviks, at least three versions of workers' control coexisted: the immediate subordination of the factory committees within a centrally directed economy; the establishment of an independent national hierarchy of factory committees; and the transfer of state power to the soviets and the creation of a workers' control apparatus within that state structure.[48]

Political exigencies did not allow the Bolsheviks to define their positions much more clearly in the second half of the year. In lieu of adopting a new economic program, the Sixth Party Congress (July 26–August 3, 1917) passed only a series of resolutions for the transition ahead. These included extending workers' control, restoring urban–rural economic exchange, and placing the banks and existing industrial syndicates under the supervision of workers' committees, which prerevolutionary owners and managers would continue to direct until the workers acquired the requisite expertise to assume full management.[49] In September and October, Lenin publicized the main elements of state capitalism, his program for the transition to socialism: the state regulation of industry; the nationalization of major syndicates; and the enlistment of existing experts into the service of the revolutionary state. Using the wartime economy of Europe as his model, he advocated compulsory syndication of industry under a watchful proletarian bureaucracy.[50] Capitalism would thus provide not only the organizational framework of the socialist economy in the form of exising monopolies but also the expertise to guide the emerging workers' state through the transition. The Bolsheviks clearly did not fear bureaucracy per se but the bourgeois domination of the existing apparatus. They sought not to eliminate bureaucracy but to convert it to a higher purpose by changing its class orientation to proletarian. As presented by Lenin, the transition to socialism, although gradual, would be relatively uncomplicated. State power would rest with the working class, who would utilize all existing resources to their own class ends.[51]

These proposals did not foreshadow the subsequent centralization of the Soviet economy as much as an initial reading might indicate. Lenin and others used the expressions "nationalization" and "plan" freely, but in context these connoted coordinated public supervision and economic rationalization rather than the centralization of economic policymaking. State capitalism was, in fact, fundamentally fragile from the outset: a centralized coordination of the economy was to be achieved, but success depended on the active participation of workers' representatives at the various administrative levels. Hence, leading Bolshevik pronouncements on the economic transformation may have stressed the role of central institutions,[52] but any realistic chance of efficacious implementation lay in the hands of regional and local activists as well as, of course, the mass workers themselves.

This factor was central, since the Bolsheviks lacked the power and means simply to impose their will. They possessed no extensive national apparatus, despite a numerical growth from 23,600 to 115,000 members between February and October and a rise to 390,000 by March 1918. By December 1917, the Bolshevik party had established provincial committees in only nineteen of the fifty-seven provinces, and in the 472 districts (*uezdy*) and towns of those provinces there were but 141 town and 2 district committees.[53] Party organization became a high priority thereafter, and between mid-1918 and the Eighth Party Congress of March 1919 party cells extended to every rural district, while party operations as a whole became more centralized.[54] These gains were mitigated, however, by the fact that the executive committee of the soviet, not the party cell, remained the operative local institution, in no small measure because the executive committees financed party operations. As late as 1919, local party committees had no full-time staff, and one Central Committee secretary disparagingly referred to them as "the agitation departments of the local soviets."[55] Moreover, as late as March 1919 the Central Committee had no systematic contact with half of its district organizations.[56] In the short-term, therefore, party effectiveness depended on establishing a Bolshevik presence within the soviet executive committees, and this is where their power lay. Their degree of success can be measured indirectly by the fact that Bolsheviks accounted for but 51 percent of the representation to the Second Congress of Soviets in October 1917, 66 percent at the Fifth Congress of Soviets in June 1918, and 97 percent by November 1918.[57]

Party statistics alone, of course, cannot tell the whole story. The disappearance of the non-Bolshevik socialists from the soviets by November 1918 reflected to a large degree the recruitment of many previous non-Bolsheviks into the party. The influx of large numbers of new rank-and-file members also attested to the fact that the meaning of party membership was beginning rapidly to gravitate away from the ideological orientation and revolutionary commitment of the pre-1917 cadre. Hence, not only did the party lack the apparatus to impose itself on society. It also faced the task of redefining its goals and orientations from within. The appearance of so many new party recruits of uncertain quality at this time obviously complicated such tasks. In these circumstances, the ability to influence local priorities into different channels was minimal, and the primacy of local aspirations continued to prevail.

Politicization and the Strike Movement

The conflation of the conditions in the textile industry with nascent Bolshevik support in the Central Industrial Region provides a metaphor for the revolution itself. After unpromising beginnings in February, the textile workers became a force in the working-class politics of the region in 1917, largely because they developed a functioning movement outside Moscow. Nothing, in fact, demonstrates the textile workers' emergence more forcefully than the development and execution of the Ivanovo–Kineshma textile strike (October 21–November 17, 1917), particularly since factory committees and regional textile union activists oversaw much of its preparation. Thus, although largely unorganized at the beginning of the year, more than 300,000 textile workers from 114 enterprises in Vladimir and Kostroma provinces would participate in one of Russia's most highly coordinated and significant labor actions of 1917 on the very eve of the October Revolution.

Few anticipated the Ivanovo–Kineshma textile strike, at least in the form it assumed. As we have seen, the workers' propensity to act was never in doubt. The important part the Ivanovo–Kineshma workers played in the recurrent major textile strikes earned them a strong revolutionary reputation well before 1917,[58] but their ability to maintain their efforts, as already noted, inspired less confidence. The textile workers' sporadic volatility failed to crystallize into a sustained and coordinated movement, even by the February Revolution. Workers' attitudes in localities where force was used against strikers had hardened during World War I, to be sure, but leaders of the textile workers' union nevertheless conceded that no viable professional organizations existed in the 280 largest enterprises of Ivanovo–Kineshma in February 1917. They considered the 100 that organized hastily thereafter to be ineffective and lacking coordination.[59] Even by late summer, neither the Ivanovo–Kineshma Union of Textile Workers nor any other working-class institution appeared capable of leading a strike on the scale of the entire region.

The factory owners were much better organized. Entrepreneurs benefited from long-term business affiliations and mutual experiences on government regulatory commissions, and during World War I they had established more than 175 industrialists' organizations nationally.[60] The Society of Factory and Works Owners (*Obshchestvo fabrikantov i zavodchikov*), created in 1906, had branches dominated by textile proprietors in Moscow and eight surrounding provinces by 1917.[61] On March 19–22, P. P. Riabushinskii chaired the First All-Russian Trade–Industrial Congress in Moscow, which led to the formation of the All-Russian Union of Trade and Industry.[62] On May 9, A. M. Neviadomskii and S. A. Morozov led fellow textile factory proprietors of the Central Industrial Region to form their own Union of the Unified Industry (*Soiuz ob "edinennoi promyshlennosti*).[63] Even these rudimentary organizations proved sufficient to reject decisively the demands for higher pay and improved conditions that the initial conference of workers' representatives of Ivanovo–Kineshma presented to owners on May 10–12.[64] In mid-June, the Provisional Government empowered major manufacturers to organize a national regulatory organ, Centro-Cloth (*Tsentrotkan'*), which the workers viewed as the owners' instrument for manipulating prices.[65] In general, factory proprietors repeatedly cited the modest terms of the Provisional Government's statute on factory committees[66] to protect their hegemony over management prerogatives and labor conditions.

The absence of coordinated regional organizations did not prevent the textile workers from attempting to redress grievances at the factory level, however, and conditions in individual Ivanovo–Kineshma mills were as volatile as any in Russia. S. A. Smirnov withheld materials in April and closed his Likino Mill (Pokrovskii district, Vladimir province) completely on September 2 rather than grant the eight-hour workday that his workers had begun to demand in March.[67] In the Ivanovo–Voznesensk Weaving Mill, the factory committee dismissed the directors of the spinning and weaving operations on May 31 and established its own management. The Union of the Unified Industry complained that this was no isolated incident,[68] and such scenarios indeed repeated themselves endlessly at the local level. Countless factories in the region initiated independent economic strikes,[69] and workers' representatives everywhere pressed for a greater voice in decision making. Throughout the summer, local soviets and factory committees would continue to pass innumerable ad hoc resolutions on mill closings, supply and distribution problems, and disputes over management authority and wages.[70]

In effect, dual power existed within virtually every industrial enterprise. In the aftermath of the February Revolution, general workers' meetings elected a factory committee to represent them before management. Although not uniform, these usually included a chairman, vice-chairman, and secretary as well as an unspecified number of additional members, depending on the size of the enterprise. Factory committees, in turn, would either attempt to establish workers' control themselves or create a control commission (*kontrol'no-khoziaistvennyi komitet*) to do so. Many factory owners resisted vigorously, principally by withholding materials and initiating lockouts. *Tekstil'nyi rabochii*, the journal of the Union of Textile Workers of the Central Industrial Region, reported extensive closings of textile plants in August–September 1917 both within Moscow and outside the city,[71] and by one count thirty-three of ninety-seven Moscow textile enterprises surveyed in June, mostly large-scale plants, had closed by October.[72] While there are no reliable, detailed, and complete figures for closings in the Central Industrial Region, one source estimates that on a national scale 20 percent of the textile mills closed by the time of the October Revolution.[73] Given the supply and transportation crisis that underlay the fall of tsarism, one cannot attribute this trend only to owners' intransigence, but we can say with assurance that factory committee and union activists perceived the closings almost entirely as the product of management duplicity.[74] In an early attempt at a corrective, the Union of Textile Workers created a special commission in Moscow on June 19 to transfer extant fuel and materials from closed factories to others in need,[75] and at the beginning of July the Ivanovo–Voznesensk Soviet endorsed this practice.[76]

Factory owners publicly labeled factory committee and union operations as anarchy, a charge that chose either to deny the legitimacy of any worker activism or to equate whatever excesses that occurred with the actual goals of the movement. In reality, the activists' efforts possessed an internal consistency, although this was not admitted or always even recognized by their adversaries. Indeed, when a regional conference of factory committees created the Ivanovo–Kineshma Union of Textile Workers in mid-June, union leaders from the outset consciously adopted a dual tactic: to organize resistance on the scale of the entire region and to link organizational efforts to the wage demands relentlessly arising from the factories.[77] Union activists then spent two months gathering data that showed nominal wages in the textile plants were at best half of those in other industries and real wages were declining precipitously. Given the particularly bitter battles taking place since February in Orekhovo–Zuevo and throughout Ivanovo–Kineshma over the issue of real wages,[78] these findings hardly caused surprise. The aim, however, was not to confirm the existence of an obvious crisis but to organize local unions around the wage campaign as the first step toward a strike the union leaders considered inevitable.[79]

Responses in the Ivanovo–Kineshma factories were as diverse as the structure of the industry and as varied as local circumstances. Large-scale factories were common but unevenly distributed. Ivanovo–Voznesensk contained sixty large-scale textile enterprises.[80] Clusters of factories located in smaller towns and industrial villages, which became the operating centers of intermediate-level union sections and strike committees, constituted a second important pattern of organization within the region.[81] Finally, there existed individual plants attached to a single village or remote area, often operating in isolation. In such enterprises, activism frequently depended on the arrival of agitators from outside. Political consciousness was low, and familiarity with national issues there was often negligible. The question of party affiliation, when it surfaced, was a topic mired in confusion,[82] and one is struck above all with how seldom the issue appears in the record. In these conditions, union organizers were greeted with enthusiasm in some localities but faced violence in others, as when union leaders A. N. Asatkin and G. K. Korolev were nearly beaten while agitating at the Kamenka Factory near Kineshma.[83] Setbacks notwithstanding and with wage information in hand, a union conference of August 15–16 responded to the priorities encountered in the factories by creating a twenty-two-member commission composed of rank-and-file workers to formulate a uniform minimum pay rate, a *tarif*, for the region.[84] When this commission presented a demand for a 7.50-ruble daily minimum on September 11, union representatives immediately approved it and circulated the proposal among the factories for discussion. On the same day, the union created a twenty-four-member strike committee to act in the event the demand were not met.[85]

While these developments were taking place in Ivanovo–Kineshma, Bolshevik activists in Moscow worked to revive the moribund Union of Textile Workers of the Central Industrial Region as the nucleus of a national union. The two-story house on Nemetskaia Street, in which the Bolsheviks I. I. Kutuzov (chairman), Ia. E. Rudzutak (secretary), and M. V. Rykunov set up quarters just after the February Revolution, at first served more as a rallying point and distribution center for radical literature than as a command post.[86] Tangible organizational successes did not appear until early June, when the factory committees of five major Moscow textile enterprises—Tsindel', Prokhorov, Danilov, Giubner, and Riabov—sent representatives to the Tsindel' Factory to discuss the economic dislocation and lockouts and to endorse workers' control. Then, on June 16–17, the leadership of the Union of Textile Workers followed by gathering the representatives of sixty-four factory committees of the Central Industrial Region to pass resolutions similar to those of the Tsindel' meeting.[87] By the time the Third All-Russian Conference of Trade Unions met in Petrograd on June 20–28, the textile union grew to 178,560 members—second in size only to the 400,000-strong Union of Metal Workers[88]—even though new enrollments raced well ahead of the emergence of an authoritative national union leadership. On September 3, some 300 delegates representing 175,000 textile workers gathered at the Tsindel' Factory again to hear reports on deteriorating economic conditions and to pass unanimously a resolution advocating workers' control. The Moscow union leadership began to collect a strike fund three days later.[89]

At this juncture, the Moscow Union of Textile Workers attempted to absorb the Ivanovo–Kineshma preparations into their own plans for a textile strike of the whole Central Industrial Region. Discussions of the establishment of a national union and of economic grievances dominated proceedings when twenty-six union sections, including Ivanovo–Kineshma, sent eighty-three delegates to Moscow on September 23–28 for the First All-Russian Conference of Textile Workers. A consensus considered an industry-wide strike unavoidable, and the leaders of the Moscow union, declaring their followers ready to respond "to the first call of a strike committee," made no secret of their eagerness to assume its leadership. The conference could not, however, resolve the question of local autonomy either in connection with the creation of a national organization or regarding the strike itself.[90] No specific plan for action emerged.

In the weeks following this conference, therefore, worker activists in Ivanovo–Kineshma continued strike preparations independently, even though they were still unsure of their full influence at the factory level. On October 7, a conference of factory committee representatives from Ivanovo–Voznesensk reviewed the discussions of the 7.50-ruble daily minimum that had been held in individual enterprises and endorsed the demand themselves. The delegates also established a timetable for action: a combined meeting of all working-class institutions on October 11; the communication of demands to the owners on October 12; and the designation of October 18 as the deadline for a reply. The delegates then instructed each factory to elect a five-member strike committee and each area to create intermediate committees.[91]

Subsequent developments followed this agenda closely. While the more than 1,000 union and factory committee activists whom the Moscow union gathered on October 9 voted to defer a decision on a general textile strike until October 18,[92] representatives of the local unions, factory committees, and soviets of Ivanovo–Kineshma met as scheduled on October 11 in Ivanovo–Voznesensk. They invested a streamlined nine-member Central Strike Committee with full discretionary power to initiate a walkout and elected a commission to negotiate with owners as appropriate. The conference sent demands for higher pay and improved sanitary and living conditions to the proprietors on the following day,[93] and the Central Strike Committee publicly directed the workers to take "neither a single forward nor backward step without the instructions of the Central Strike Committee."[94] When management failed to respond to demands by October 18, the committee ordered the factories to strike and, still uncertain of the will and ability to comply, sent representatives to twelve main textile centers to oversee final preparations. Even within Ivanovo–Voznesensk, city and factory strike committees deemed it necessary to issue precise instructions on topics ranging from the use of arms and the prevention of looting to the timing of the singing of the Marseillaise. Three days later, the long-anticipated work stoppage became a reality.[95]

Nevertheless, as the Ivanovo–Kineshma activists were finalizing their preparations, they fell increasingly out of synchronization with working-class politics in Moscow, where workers' organizations rejected labor actions in anticipation of the imminent political revolution. Union leaders and very possibly skilled workers in Moscow began actively discouraging strikes in mid-October.[96] On October 15, a meeting of the Moscow Union of Textile Workers condemned the owners' intransigence in wage negotiations[97] but passed a resolution on the necessity of presenting a common front through the Central Bureau of Trade Unions and with other revolutionary organizations in lieu of independent actions.[98] Then, on the day following the beginning of the Ivanovo–Kineshma strike, 110,000 Moscow leather workers ended a nine-week walkout. Finally, although the Ivanovo–Kineshma strike would spread into Moscow province as far as Bogorodsk and Pavlovskii–Posad, the Moscow textile union failed to implement the strike plans it had proposed in September.[99]

In Ivanovo–Kineshma, labor and management immediately assumed assertive public postures once the walkout began. On the initial day, the Central Strike Committee issued a series of three declarations that restated their grievances, called for discipline, and solicited support from soldiers.[100] The owners publicly condemned the strike as "anarchic"[101] while urging in private communications that government authorities use force against the strikers. On October 21, Neviadomskii, speaking as chairman of the Union of the Unified Industry, demanded "decisive and immediate" state intervention from A. I. Konovalov, Minister of Trade and Industry and proprietor of textile mills in Ivanovo–Kineshma.[102] On the same day, the Society of Factory and Works Owners insisted that the Ministry of Trade and Industry and the Ministry of Labor "take immediate measures toward the reinstatement of the laws breached by the Strike Committee."[103]

Direct communications between labor and management were only slightly less stilted. On October 22, the Union of the Unified

Industry proposed to the Strike Committee that negotiations be scheduled. The industrialists, in fact, claimed to have answered the strikers' demands on October 14 and offered the preposterous suggestion that the response was lost in transit. The offer did not, however, include the implementation of the 7.50-ruble *tarif* as a precondition to talks, a key worker demand. Suspecting a ploy to split the strikers' ranks, the Central Strike Committee instructed the local areas to stand firm. On October 24, the Menshevik K. A. Gvozdev, Minister of Labor, invited delegates to Moscow for talks but was told: "Your telegram is unclear. If the meeting intends to discuss the All-Russian conflict in the textile industry, then we can delegate representatives to Petrograd. If the representatives are being called to settle our own conflict, then it would be more fitting for your delegation to come to our region, in the center of the strike."[104] Just before midnight, the committee received a new offer from the Union of the Unified Industry, which included a proposal to discuss a "subsistence maximum" (*prozhitochnoi maksimum*). The Strike Committee sarcastically requested a definition of "subsistence maximum" and reiterated its 7.50-ruble wage demand.[105]

The Bolshevik seizure of political office in Petrograd on October 25–26 cut off these communications, sent owners into hiding, and— ironically—undercut the Bolshevik-dominated Central Strike Committee. Amid the general euphoria surrounding the fall of the Provisional Government, the strike leaders reached the sobering conclusion that their labor action had lost much of its logic. The owners themselves openly resolved to halt production in reaction to the revolution, and it was now the revolutionary government that desired a return to stability. For the first time, therefore, the strike movement had to face a conflict between its enmity toward the propertied classes and support for their overthrow, on the one hand, and the economic goals of its strike, on the other. There was no shortage of support in Ivanovo–Kineshma for the revolution itself. In contrast to the passivity of Moscow factory workers, local union sections throughout Ivanovo–Kineshma volunteered to aid in the seizure of power in Moscow. Although the Moscow Soviet declined the assistance, M. V. Frunze and local strike committee chairman Bogdanov led a detachment of 300 Shuia workers who were among the first to enter the Kremlin after days of armed fighting in the city.[106] The wisdom of continuing the strike under a new revolutionary regime began to be questioned even before this. As early as October 27, strikers in the key manufacturing centers of Shuia and Kovrov began to raise the question of ending the walkout.[107] The Central Council of Trade Unions placed the termination of all strikes at the top of its agenda of November 1,[108] and the Moscow Bureau of Trade Unions issued a joint resolution with the Moscow Military–Revolutionary Committee on November 4 urging a general resumption of work.[109] On November 6, the Moscow Union of Textile Workers sent a three-member commission, including a representative of its newly elected All-Russian Council, to Ivanovo–Kineshma to end the work stoppage.[110]

The strike leaders agreed in principle to return the workers to the factories but, given the economic demands from the enterprises that had provided their walkout its original impetus, felt it imperative first to achieve some tangible gain. In an effort to salvage a compromise, therefore, the Central Strike Committee sent more moderate demands to the owners on November 9.[111] On the following day and before a reply could arrive, however, the People's Commissariat of Labor instructed the Ivanovo–Kineshma activists to settle their strike and ordered the Moscow workers not to join the walkout.[112] While these events were unfolding, the Ivanovo–Kineshma strike movement encountered an additional challenge. The Central Bureau of Trade Unions began to appropriate the role of spokesman for all textile workers by orchestrating talks in Moscow among the Union of the Unified Industry, the Commissariat of Labor, and itself.[113] With their leverage thus disintegrating, the Ivanovo–Kineshma activists could attain little more from the owners in meetings held November 11–14 than a promise that wages would be renegotiated once the workers returned to the enterprises.[114] Mills reopened on November 17, and despite the terms of the settlement the workers' press reported a victory.[115]

In reality, the end of the strike marked neither victory nor defeat but the beginning of a more contentious stage in the battle for authority in the factories. Some owners tried to prevent reopenings on November 17, and lockouts increased thereafter as industrialists' organizations intensified and attempted to unify their resistance to workers' control. Textile factories and local organs consequently generated numerous reports of so-called bourgeois sabotage—lockouts, abandonment by the owner, refusal to accept workers' control—and the union press publicized their grievances.[116] In these circumstances, the vague promises of November 11–14 held little prospect of producing peace. With the workers back in the plants, the owners introduced their own interpretation of a subsistence wage: 5.25 rubles per day for men and 4.50 for women. With no agreement in sight, talks collapsed on November 29.[117]

None of these developments inclined the regional activists to relinquish any authority, however, even to superordinate working-class institutions. At the November 26 meeting of the All-Russian Council of the Union of Textile Workers, Asatkin, chairman of the Ivanovo–Kineshma union and a member of the All-Russian Council, clashed with Rudzutak, chairman of the All-Russian Council and soon to be appointed to the presidium of *VSNKh*,[118] over the issue of local autonomy. Although not couched strictly in terms of centralization versus localism, their debate nevertheless made it clear that Asatkin and his followers would support a national union at present only insofar as it was organized on the basis of small, locally oriented sections.[119] The All-Russian Council soon reprimanded the Ivanovo–Kineshma activists for continuing to conduct independent negotiations for a *tarif*.[120] When the wage negotiations then foundered on November 29, however, a caucus of radical activists from the union and the Ivanovo–Voznesensk Soviet initiated an even more direct strategy: kidnapping leading industrialists in Moscow and jailing them in Ivanovo–Voznesensk. By December 3, four were in custody, including—surely the biggest prize—Neviadomskii.[121]

Local institutions asserted themselves in other ways as well. At the beginning of December, the Ivanovo–Kineshma Soviet issued detailed instructions on the firm steps to be taken against the owners' attempts to block the implementation of the new union *tarif*. On December 2, the union advocated the organization of courses on workers' control. The soviets, factory committees, and union sections of Ivanovo–Kineshma issued a joint resolution on December 6–8 that explained the creation and operation of the control commissions that would supervise factory operations. On December 19, the Ivanovo–Kineshma Commissariat of Industry released a plan for supplying the factories with materials and fuel.[122] In most cases, these never progressed beyond the passage of a resolution but served as declarations of intention.

Even these limited independent initiatives of the Ivanovo–Kineshma activists began almost immediately to conflict with the campaign to create national working-class institutions and policies. The Decree on Workers' Control that *Sovnarkom* issued on November 14 attempted to define the tasks of the accounting and supervision of production on a national scale.[123] The creation of the Supreme Council of the National Economy on December 1, whose mandate called for the creation of a regulatory center or *glavk* for each industry, aimed at the unification of all economic life under a single institution.[124] In addition, decrees specifically designed to curtail speculation in textile goods appeared on December 8 and 24.[125] On December 16, *VSNKh* published a project of the Moscow Soviet that created its own Centro-Cloth, an institution designed to account for all textile production.[126] These national decrees, which technically superseded the lower declarations, required the subordination of some local interests to long-term goals, but the present infringement on independence of action was only implied. Questions of jurisdiction were left largely unanswered.

There can be no question, however, that the Moscow union leaders began to claim the leadership of the workers' movement in the textile industry immediately after October, especially as workers' and owners' representatives there vied to dominate the regulation of production and distribution. As we recall, the Provisional Government's attempt to create a cotton monopoly in June 1917 by commissioning industrialists to found Centro-Cloth only intensified opposition from textile factory committees and union sections. Labor had no representation in this organ, and their demands for a voice in regulating the industry before October 1917 became only more insistent thereafter.[127] Ownership, in fact, experienced pressure from various Moscow sources. On November 18, a conference of factory committees from Moscow's textile factories proposed a new system of representation in Centro-Cloth based on the proportion of owners, employees, and workers in the industry and circulated their plan to general factory meetings for discussion.[128] On November 28, following the November 26 meeting of the All-Russian Council of the Union of Textile Workers that failed to unite the regional unions into a single force,[129] the Moscow textile factory committee representatives met again and proposed that the Economic Section of the Moscow Soviet govern local industrial

production.[130] By December, as tensions in individual plants
heightened, workers began regularly to haul troublesome owners
before general factory meetings. If this tactic failed to intimidate,
they would then halt production until management recognized
workers' control. In the same weeks, the Moscow Union of Textile
Workers began to seize textile output and established three wholesale and numerous retail outlets to distribute its goods.[131] When
Centro-Cloth voiced objections, a December 7 meeting of the factory committees of Moscow's cotton mills threatened to take over
Centro-Cloth's organs of distribution.[132]

In the face of this pressure, Centro-Cloth admitted a nine-
member Workers' Group on December 10, but the implementation
of this concession only further convinced the workers' representatives of the need to increase their own authority. Employees threatened to stop work in Centro-Cloth when the workers first appeared,
and the owners largely ignored the Workers' Group. Most important, owners and employees were able to retain the so-called parity
principle, by which owners, employees, and workers voted as cu-
riae. Not surprisingly, owners and employees closed ranks to out-
vote the Workers' Group.[133] In response, the All-Russian Council of
the Union of Textile Workers and the Workers' Group of Centro-
Cloth met jointly on December 13 to devise a new strategy. Indicative of the Moscow union's growing assertiveness, the gathering
resolved to convene a national union congress to discuss the condition of the industry, regulation, workers' control, and especially the
reorganization of Centro-Cloth.[134] Within a week, the union leadership also instructed factory control commissions in textile enterprises to take full supervision of production.[135]

Thus, although the Ivanovo–Kineshma strike movement had
spearheaded textile activism in 1917, the union leadership in Moscow attempted to appropriate the main role in transforming the
industry after October. The end of the strike, however, by no means
conceded anything fundamental to superordinate organs. The return to work was more than anything a recognition that the October
Revolution had altered the political agenda. It neither signaled that
grievances had been redressed nor indicated that the government
agencies instrumental in restoring production enjoyed firm support.
When the Ivanovo–Kineshma strike movement ceased operations,
therefore, it did so quietly but without finality. In the last act

connected to the strike itself, the Ivanovo–Kineshma Union of Textile Workers unilaterally sent new wage demands to the owners on December 15. Expecting no response, it then declared itself in a state of "irreconcilable struggle." The denouement occurred in late December, when the Moscow Soviet effected the release of the captive industrialists, a solution that caused only resentment in Ivanovo–Kineshma.[136]

Conclusion

The unharmonious workers' movement and inconclusive strike within the textile industry pointed out important cleavages within the constituency of the revolution, even among those supportive or tolerant of the Bolshevik seizure of office. The local reports that reached conferences and congresses both before and after the October Revolution continually pointed out that economic demands dwarfed all other issues in the plants, a point by no means lost on union activists. Caught between their insecure base of support in the factories, pressures from superordinate working-class institutions, and the opposition of the factory owners, conscious workers in the industry attempted a delicate balance: to lead the rank and file toward greater and more coordinated political assertiveness; to represent faithfully demands rising from factories where the activists enjoyed only a tenuous influence; and to protect local freedom of action. In these circumstances, members of the soviets, factory committees, and trade unions could and would cooperate toward shared goals, but inter- and intrainstitutional rivalries were at least equally strong. The influence of the revolutionary intelligentsia, meanwhile, appeared small and indirect at the local level, and evidence of the importance of party affiliations, as we have seen, did not emerge with anything resembling the frequency of factory and local loyalties. In Ivanovo–Kineshma specifically, a high incidence of Bolshevik party membership among the strike leaders in no way guaranteed obedience to directives from above, even those of the Bolshevik-dominated leadership of the Moscow Union of Textile Workers. At the time the Petrograd Bolsheviks seized national political office, therefore, the party leadership was better situated to articulate long-term plans—many of which were only

abstractions to the unskilled or semiskilled worker at the bench—
than to establish its authority in the present. To the contemporary
observer, the experience of the textile industry of the Central Industrial Region clearly showed that conditions in existence for centuries could not be reversed only by a transfer of political power.

Activism in the center of the Russian textile industry in 1917
therefore foreshadowed the enormous dimensions of consolidating
the revolution. Resistance from the beneficiaries of the old order
was to be expected. More troubling to supporters of the revolution
was the proliferation of soviet, union, and factory committee organs attempting to perform the same functions, the intrainstitutional conflicts, and the uncertainty surrounding the mood and
proclivities of the factory workers themselves. In such circumstances, the relationship between institutional arrangements and
actual practice was mediated by contemporary events, and consequently was in a perpetual state of renegotiation. Authority, insofar
as it existed, resided with those who could best address contemporary demands, regardless of formal organizational hierarchies.
Since no organ could consistently satisfy present demands, however, the situation was one in which various institutions endlessly
asserted themselves without any actually establishing effective administration. Reorienting this industry would be a task of major
proportions.

3

The Revolution in Practice:
The First Year

The requirements of consolidating the revolution strained the unity of October, rooted in a common commitment to ending oppression, beyond its limits. The diversity of the expectations of "revolution," sublimated during the overthrow of the Provisional Government, now exerted a strongly centifugal effect. At the same time, party guidelines for the gradual construction of socialism exercised but scant influence over mass behavior and failed to produce the smooth socioeconomic transition confidently predicted in prerevolutionary rhetoric. Problems did not derive simply from ideological miscalculations. Before October 1917, the Bolsheviks were not alone in their failure to anticipate the further intensification of the economic crisis, the mass exodus from the cities, and the vicissitudes of international events in 1918. Rather, reduced to essentials, the chief conflict for the revolutionary state and party lay in reconciling the divergent short-term perceptions of "revolution" among the revolution's (largely self-selected) constituency. The quintessence of this scenario was found in the Russian textile industry, where workers and their representatives at the national, regional, and local levels exhibited far less than full compatibility in their interpretation of responsibilities.

In the most visible position, the national leadership of the textile workers' union adhered more closely than other actors to the ideal of the revolution then being articulated in party declarations and decrees, but this by no means resulted in consistent obedience to national leaders. Compliance with directives was regularly modified by the union leadership's pursuit of an additional double-edged orientation whose requirements frequently took precedence: continuing to defend class interests against the remaining owners and their representatives in the industry, and actively extending national union authority against other working-class institutions and over lower union organizations and members. To this end in 1918, the union leaders at the national level enjoyed increasing support from the officers of regional union organizations such as Ivanovo–Kineshma, who, in turn, became more visible in national union politics without fully abandoning their regional priorities.

At the opposite end of the spectrum were the rank and file, largely mass workers throughout the Central Industrial Region and in Moscow. These elements continued to define their goals as the defense of local economic interests and the subjugation of oppressive management, the dual impetus of the 1917 workers' movement in the industry. The pressure that the rank and file generated locally did not exclude support for the greater implementation of authority on a national scale, provided that the establishment of such authority led to an alleviation of local crises. Paradoxically, this impulse did not include sympathy for a surrender of local autonomy in the process.

Caught between the two poles were the (largely but not exclusively conscious) activists, that is, those who filled positions on the factory committees and control commissions as well as on the administrative institutions that came into being under Soviet rule. Impatient constituencies continually pressed these activists toward greater assertiveness and increasingly subjected them to criticism. These workers' representatives and officials found themselves simultaneously on the defensive: against jurisdictional encroachments laterally and from above; against their own inexperience and lack of expertise in the face of pressing tasks. As a result, the frequently unmanageable behavior of mass and conscious workers in the Russian textile industry did not result so much from a disillusionment with Bolshevism as from the projection of variegated perceptions of "revolution" onto its 1918 agenda.

All of this undermined administration. National directives repeatedly outlined a theoretical hierarchy of institutions and formulas for cooperation among organs. We have already seen, however, that this produced more duplication of effort and jurisdictional disputes than order, and the full issue goes even beyond this. Above all, the conditons of the consolidation of the revolution created an atmosphere in which no relationship or issue could be considered resolved with finality. Consequently, institutional operations not only lacked order, but in such an environment each organ continually tried to enhance its position regardless of the authority formally devolved upon it. In the textile industry, this type of continual struggle emerged in 1918 between the representatives of management and workers in Centro-Cloth and later Centro-Textile, as well as among union, *sovnarkhoz*, soviet, and factory organs at each administrative level.

The Beginnings of the Soviet State, 1918

Multifarious pressures on the Bolsheviks mounted literally from the moment of the October Revolution. Several days of bloody skirmishing outside Petrograd followed the seizure of office. Intense fighting erupted in Moscow as well before the revolutionaries consolidated their control of the cities. National minorities took the Bolshevik advocacy of self-determination at face value and in short order began to break off from the empire. Within the capitals, the Bolsheviks faced the refusal of civil servants to serve the revolutionary regime. This they overcame only with great difficulty, and in the most severe cases strikes of office workers and teachers extended into the spring of the following year. At the Second All-Russian Congress of Soviets, convened the day following the seizure of office, the Mensheviks and right wing of the Socialist-Revolutionaries (SRs) walked out, leaving the Bolsheviks the opportunity to pass highly popular decrees on peace and land as their first two legislative acts. This did not, however, eliminate the considerable sympathy for the formation of a socialist coalition government, a sentiment shared even by moderate Bolsheviks. Consequently, the All-Russian Central Committee of the Union of Railroad Workers (*Vikzhel*) used its considerable leverage to force the issue by threatening a general strike. This was averted only when Lenin agreed to negotiations that began four days later, and Left SRs agreed to enter the Council of People's Commissars in early December. Elections for the long-awaited Constituent Assembly went ahead as scheduled in November, with the SRs receiving the largest portion of the votes and the Bolsheviks garnering but one-fourth of the ballots. The Assembly met for only one session in January 1918 before prorogation.

Meanwhile, the economy continued to elude direction. The ruling Council of People's Commissars created *VSNKh* at the beginning of December and directed it to establish a regulatory center (*glavk*) for each sphere of industry, but this provided only future solutions to immediate crises. The flight of a large portion of the country's administrative and economic experts in the months following October did not help this situation, nor did the wide-scale reassignment of the factory activists of 1917 to other tasks in the first months of Soviet rule. Meanwhile, the deepening famine and increasing unemployment began to reduce the urban population to a fraction of its former size. Ration cards appeared in April 1918, followed by a series of declarations against hunger in the final days of May. The formation of Committees of the Poor (*kombedy*) on June 11, an attempt to exploit the animosities of the poorer peasants and mobilize them against those better situated, failed to alleviate the grain shortage. By mid-October 1918, administrative measures against famine included the institution of four categories of bread rations but had little impact on curtailing widespread hunger.

Developments in the international arena also brought little cause for optimism. Party loyalists of all factions eagerly and then with growing anxiety awaited the international proletarian revolution on which they counted for the material support and expertise required by their modernizing revolution in a devastated, agrarian country. Nor did the Bolsheviks' attempt to forge a new diplomacy succeed. The Allies ignored the Decree of Peace issued October 26, and the Central Powers took it only as a sign to increase pressure. On December 2, the Central Powers signed an armistice with the revolutionary regime, but it was months before this produced an actual peace treaty. It did not help, in all candor, that Leon Trotsky was named Commissar of Foreign Affairs. "What diplomatic work will we have?" he asked rhetorically. "We will publish a few revolutionary proclamations to the people and then shut up shop."[1] Such thinking gave rise to serious miscalculations. Trotsky assumed that all of Europe was on the brink of revolt and feared that making peace might calm radical ardor and delay the international revolution. When the Central Powers demanded that Russia give up Poland, Lithuania, and western Latvia—territory the Germans already occupied—Trotsky therefore refused. Instead, in a flamboyant gesture of revolutionary theater, he announced to the German negotiators that the Soviet Republic would pursue a policy of "no war, no peace" unilaterally. This, coupled with the Soviet Republic's formal demobilization on January 29 of its already disintegrating army, presented the Germans with an unprecedented diplomatic strategem. The Germans did not remain nonplussed for long. Eight days later, on February 18, they resumed a state of war, advancing on Russia's western boundary.

At this stage, Lenin assumed personal responsibility, and his solution precipitated the most serious challenge to his authority within the party. He sponsored a treaty signed at Brest–Litovsk on March 3 in which the Bolsheviks gave up not only Poland, Lithuania, and western Latvia, but the Ukraine, Estonia, Finland, and the Transcaucasus. Opposition to the treaty within the party and ruling coalition was widespread and vehement. The Left SRs withdrew from the governing coalition in protest, and divisions among the Bolsheviks over the treaty escalated into a questioning of Lenin's gradual tempo for building socialism. In addition to opposing acceptance of the harsh terms of Brest–Litovsk, the Left Communist faction vigorously campaigned for an accelerated exclusion of the former owners and managers from economic administration. As tensions peaked, Lenin threatened to resign from the party Central Committee and *Sovnarkom* if his position were not accepted, and he ultimately prevailed. The regime would annul the Treaty of Brest–Litovsk on November 13, but by then it would be fully engulfed in another struggle.

The Bolsheviks did not enjoy the leisure to resolve their problems methodically, but did so under the direct threat of the disintegration of the republic itself. Fearing German military pressure, the government evacuated Petrograd, and Moscow formally became the capital on March 12. Internal military threats appeared as well. By the end of February 1918, General Lazar Kornilov had regrouped in the south and began an assault on territory held by the Soviet government. Although unsuccessful—Kornilov was killed April 13 and his forces retreated under Anton Denikin—the campaign foreshadowed future, more serious challenges. Antirevolutionary White armies began to take shape in the north under General E. K. Miller, in the northwest under General N. N. Iudenich, in Siberia under Admiral A. V. Kolchak, and in the south under Denikin, with the ambitious Baron Wrangel under his command. The revolt of Czechoslovakian legions in May–June 1918 as they moved eastward to leave the country marked the escalation of these limited hostilities into a full-scale civil war.

The White Armies, enjoying heterogeneous anti-Bolshevik support, gained a strong foothold in the campaigns of the summer and fall of 1918. Forces under Admiral Kolchak advanced threateningly from Siberia, the army of General Denikin captured the Ukraine and moved northward toward Moscow, and troops under General Iudenich marched within thirty miles of Petrograd. The Allied intervention brought soldiers from no fewer than fourteen countries to Russian soil by the end of the year. Already scarce supplies dwindled, and military mobilizations conscripted important human resources. Bolshevik responses were largely ad hoc. Trotsky directed the rapid organization of the Red Army; factories recently converted to peacetime use were hastily remobilized for war production. None of this, of course, promoted stability. On the contrary, the military emergency had predictably disruptive consequences that only exacerbated the problems already extant.

Workers' Control, Parallelism, and the Nationalization of Industry

In the first months of Soviet rule, central institutions found themselves in a position only to affirm their objectives in industry, not to enforce their authority. The *Sovnarkom* Decree on Workers' Control, to cite the main example, specifically empowered workers' representatives to examine the accounts and correspondence of an enterprise (point 7); to impose decisions of factory committees and control commissions on the owners (point 8); and to have the right to supervise (*nabliudenie*) production, establish output minimums, and collaborate in setting prices (point 10).[2] This was not, however, backed by authority sufficient to reconcile the competing interpretations of workers' control discussed in the previous chapter and therefore did not impose any workable degree of national coordination.

In this regard, the chief problem lay not in an absence of working-class institutions willing to implement local economic administration but in a proliferation of organs attempting identical functions, with no higher authority capable of defining jurisdictions and mediating disputes. At the center of controversy stood the factory committees, initially the most reliable gauge of working-class opinion in 1917 but increasingly challenged in their authority as the soviets and trade unions, as well as party organizations, became better organized. The Third All-Russian Conference of Trade Unions in June 1917 had, in fact, accepted the hegemony of the factory committees over the unions, but sentiment changed in the second half of the year, when the absorbtion of the factory committees by the trade unions became a recurrent refrain in the working-class politics of Petrograd. Bolshevik and Menshevik trade unionists had joined forces at the First All-Russian Conference of Factory Committees held in Petrograd October 7–12 to block the creation of an independent factory committee organization and to pass a resolution that declared the All-Russian Council of Factory Committees to be a division of the All-Russian Council of Trade Unions. The First All-Russian Congress of Trade Unions, which met in January 1918, forced the issue, and the Sixth Conference of Factory Committees, which convened immediately following the trade union congress, voted to accept their subordination to the unions. Outside Petrograd, the question was less pressing. Economic hardships and shortages of conscious workers there had forced closer cooperation among working-class institutions than in the capital, and numerous local precedents already existed for the assimilation of factory committees into the unions.

Nevertheless, a declaration from Petrograd was insufficient to guarantee the complete integration of work, and parallelism—with its duplication of effort and institutional rivalries—persisted as both a central and a local problem. Critics noted that at the highest administrative level the Central Council of Factory Committees and the central councils of the various trade unions addressed the same needs. At the regional level, the unions' attempts to handle labor questions clashed with the soviets' plans to direct industrial production. In the local areas, fluid institutional and political loyalties fostered instability. In practice, supposedly separable labor and production issues proved inextricable, and questions of jurisdiction became constant sources of disagreement among institutions at all levels.[3] This was particularly pronounced in the frequent clashes between the economic organs of the soviets and the local councils of the national economy[4] but manifested itself in other institutional conflicts as well. In one particularly pointed example in June, Rudzutak, the outspoken supporter of a centralized textile union, complained that union control commissions were in no way connected to *VSNKh* control commissions, centers and *glavki* were not fully subordinated to *VSNKh*, and sections of the textile union exercised authority over technical questions independently of the industry's regulatory organ, Centro-Textile.[5]

Reconciling the competing perceptions of workers' control proved even more troublesome. The central issue after October, as William Rosenberg has shown, was not Bolshevik state power versus the exercise of authority at the factory level but the need for central organs and the factory committees to reach a mutual understanding on the extent of local prerogatives. The First All-Russian Conference of Factory Committees had voted to expand workers' control into gradual but total regulation of the economy through a state system of unions, soviets, and factory committees. Neither this vague mandate, however, nor the *Sovnarkom* Decree on Workers' Control of November 14, which failed to resolve questions of jurisdiction, settled the issue. To complicate matters, the *Sovnarkom* decree did establish an All-Russian Council on Workers' Control that satisfied the recommendations of the October factory committee conference, but this body met only twice. When *Sovnarkom* created *VSNKh* on December 1, the new organ absorbed the Council on Workers' Control and, by extension, linked the future of workers' control to the ability to *VSNKh* to assume full direction of the national economy.[6] *VSNKh* did not accomplish this in 1918.

Instead, intense social antagonisms, the expectation of an immediate improvement in material conditions, and unbridled economic warfare overshadowed economic coordination. In response to factory proprietors' resistance or flight as well as to a desire to redress past grievances, workers and their representatives seized large numbers of enterprises between October 1917 and February–March 1918, a period Lenin euphemistically labeled the Red Guard Attack on Capital. By mid-1918, hostility between ownership and the work force provoked approximately 500 factory expropriations in Soviet

Russia, with local institutions carrying out 70–80 percent on their own authority.[7] Activity in the textile industry conformed to this pattern. While textile plants were among the first nationalized by *Sovnarkom*,[8] the state recorded only seven official nationalizations by March 1918 and but eighteen by November.[9] Actual confiscations were undoubtedly much higher. The infrequent reports that reached central organs and regular journalistic accounts consistently described the intensity of conflicts in the textile factories throughout the year[10] so that—in view of the poor communications in the industry—unrecorded textile expropriations surely exceeded official figures by a wide margin.[11]

Order, in fact, decreased more quickly than new authority emerged. Confronted by what it considered an excessive number of confiscations carried out on local initiative, *Sovnarkom* decreed on January 19 that only *VSNKh* could authorize the nationalization of factories, and *VSNKh* issued a similar declaration on February 16. The fact that *VSNKh* then found it necessary on April 27 to notify local soviets and *sovnarkhozy* that it would withhold financial support from concerns nationalized without its approval indicates how little compliance these decrees commanded.[12]

By spring 1918, Lenin's writings assumed a less confident tone. He wrote that

> we have *not yet* [late April 1918] introduced accounting and control in those enterprises and in those branches and fields of the economy which we have taken away from the bourgeoisie; and without this there can be no thought of achieving the second and equally essential condition for introducing socialism, raising the productivity of labor on a national scale. . . .[13]

> If we decide to continue to expropriate capital at the same rate at which we have been doing it up to now, we should certainly suffer defeat, because our work of organizing proletarian accounting and control has obviously—obviously to any thinking person—fallen behind the work of directly "expropriating the expropriators."[14]

Accounting, raising labor productivity, and reinstituting factory discipline all received greater emphasis in his economic pronouncements at this time.

With "expropriating the expropriators" succeeding too well, Lenin and his supporters in the party leadership turned to other tactics early in 1918, including a controversial attempt to form jointly owned industrial monopolies. Following unsuccessful negotiations with foreign capitalists in February to create ventures in which public and private capital would share ownership, the state opened talks in March with Alexei Meshcherskii, who spoke for a group of Russian and foreign entrepreneurs interested in a metal-fabricating monopoly. Bargaining reached the point that Lenin personally represented the state on April 11, but the Meshcherskii group's demand for two-thirds of the stock brought the talks to an end three days later.[15] Representatives of *Sovnarkom* conducted similar meetings with the so-called Stakheev group, another consortium of factory owners and financiers, but abandoned the project amid charges that the industrialists did not bargain in good faith. In the textile industry, tentative attempts to create jointly run factories yielded no important results.[16]

Exploring another alternative, Lenin and those who supported his position considered using the nationalization of industry—which, it should be stressed, in the spring of 1918 still connoted instituting state direction only after a regulatory apparatus was in place—as a way to increase the economic authority of central institutions. As the economist V. P. Miliutin summarized matters for the *VSNKh* meeting of March 19, the state found itself financing both nationalized and private firms at the time and needed full nationalization as a means to ensure management authority over factories for which it had already taken economic responsibility.[17] In its next issue, *Narodnoe khoziaistvo*, the official *VSNKh* journal, advocated the extension of nationalization from individual firms to whole spheres of industry as a precondition to improving the distribution of products.[18] In keeping with his concern for labor discipline and productivity, Lenin then advocated broad nationalization as part of a "categorical and ruthless struggle against the syndicalist and chaotic attitude of the [presently] nationalized enterprises."[19] While not abandoning the program of state capitalism, Lenin and his followers began to refine its implementation.

Lenin was by no means the only influential voice in the party, however, and by the spring of 1918 the Left Communist faction pressed an alternative policy. To the Left Communists, the strong reliance on so-called bourgeois personnel under state capitalism would forever preclude the emergence of a proletarian economy,

given existing class animosities, and they argued for rapidly eliminating the bourgeoisie from economic regulation. In May, Lenin
countered this challenge to state capitalism by asserting "that state
capitalism would be a *step forward* compared to the present state of
affairs,"[20] but the Left Communists remained unconvinced. Responding for the opposition, Nikolai Bukharin and N. Osinskii,
who had resigned the chairmanship of *VSNKh* in protest of the
Brest–Litovsk treaty, reiterated that their reservations were not
based on an advocacy of a rapid implementation of workers' management per se but on a deep skepticism that the proletariat and
bourgeoisie could work jointly in administration.[21] Later in the
year, Osinskii would denounce state capitalism as "a higher, mature
form of finance capital."[22]

With this issue unresolved, *Sovnarkom* and *VSNKh* continued
to extend nationalization. On May 2, *Sovnarkom* issued a nationalization decree for the sugar industry, the first full sphere of
production so affected.[23] Lenin then reiterated the need for full and
rapid nationalization at a special *VSNKh* conference held May 12–
18 to discuss the future of the metallurgy industry,[24] and the May 23
agenda of the Council of the Union of Textile Workers included the
immediate preparation of a congress to discuss the prospects for
nationalizing the textile industry.[25] On June 2, while Miliutin and
Osinskii continued to debate their respective positions on nationalization and economic administration at the First All-Russian Congress of Councils of the National Economy (May 26–June 4, 1918),
Sovnarkom decreed the petroleum industry nationalized.[26]

The general *Sovnarkom* Decree on Nationalization of June 28,
however, was neither the logical extension of these actions and
debates nor a sign that the requisite administrative organs had been
successfully created, but the result of a threat of German intervention in the Soviet economy. By early June, it became increasingly
evident that *Sovnarkom* could not ignore this German dimension in
the battle for the direction of industry. Since the signing of the
Treaty of Brest–Litovsk, the Soviet government had feared German
demands for the protection of German property, and it was particularly concerned that enterprises under German ownership would be
lost to any future nationalization project. Russian factory owners
shared this assessment of the situation. To protect their holdings
from seizure, therefore, industrialists increasingly transferred pro-

prietorship to German citizens and other foreigners, even if only on paper. When Russo-German economic talks reopened at the beginning of June, consequently, the issue of the nationalization of factories was at the forefront, and the Germans left no doubt about their position. On June 10, the Russo-German economic conference created a Subcommittee on Nationalization, and from the outset the German negotiators made sure that the transfer of factory ownership received prevailing attention. By June 20, the Germans had made it clear that they would demand compensation for all firms confiscated after July 1, and Iurii Larin communicated this to Lenin on the following day. M. B. Krasin, another Bolshevik negotiator, directly urged *Sovnarkom* to issue a general nationalization decree not later than June 29 in order to preserve Russian enterprises from German claims.[27]

Fortunately for *Sovnarkom*, the First All-Russian Congress of Councils of the National Economy had already begun preliminary work on a general nationalization decree. It had put forward preliminary suggestions on the scope and wording of such a decree,[28] and not later than June 3 Lenin personally took part in a special commission elected by the congress to work out particulars.[29] When the Soviet leadership met sometime after eight o'clock on the evening of June 27, therefore, much of its work had already been done, and *Sovnarkom*, in reality, could add little. Those present conceded that *glavki* and centers did not yet exist in all major spheres of production and that many of those in operation did not even know the number of enterprises within their respective industry. Unable, consequently, to list firms by name, *Sovnarkom* issued a nationalization order on the basis of the total capital of each factory (from 200,000 to 1 million rubles, depending on the industry).[30]

In the prevailing circumstances, the nationalization decree was a paper measure that exercised little immediate influence at the center and even less in the local areas. During the Red Guard Attack on Capital, complaints of separatism and localism in the workers' organizations of the textile-producing provinces had occupied a central place in the press and in institutional communications. These continued to appear regularly after the nationalization decree, as when the issue surfaced at the Moscow *Oblast'* Conference of Textile Workers on August 12.[31] We have also seen that Lenin criticized the chaos in factories already nationalized by the spring of 1918. Evidence that factory committees in nationalized enterprises continued to function without ties to the center also surfaced after the *Sovnarkom* decree, as at a conference of the factory committees and the representatives of local workers' organizations of Serpukhov *raion* on October 10.[32] Bemoaning the strong rural ties of textile workers, an additional leitmotif in discussions of the failure of organizational efforts, survived as well, and this factor should not be underestimated. When the textile factory committees and union officials of Tambov *raion* met August 1–3, the delegate Troitskii, who represented the Zaitsev Factory, linked organizational failures to the fact that "50% of the workers in the factories are peasants." Each of the following six rapporteurs repeated that the peasant or semipeasant character of the work force was a central problem.[33] In a comprehensive assessment of the situation, *Narodnoe khoziaistvo* criticized the "insufficient ties of the center to the local areas" and "the absence of correct central supervision of the work in the local areas," as chaotic provinces awaited and in some cases actively solicited central direction that was not forthcoming.[34]

Remedies, however, generally only repeated earlier developments. These encompassed, above all, calls for greater local autonomy[35] and a proliferation of institutions attempting identical functions.[36] At the time, the Left Communists praised the nationalizaton decree, asserting that productivity rose in nationalized plants and agitating on this basis for even greater independence for factory organs. In mid-July, however, Miliutin called their examples "not typical," argued that the reverse was more often true, and declared that the real organization of the economy still lay "in the future."[37]

The balance between short- and long-term objectives in such conditions was fragile. In order to revive exchange, one precondition common to all prescriptions for the future, it would be necessary to produce manufactured goods in significant quantities and to deliver a large proportion of them to consumers. Failing this, not even forced grain requisitions and other emergency measures, as matters turned out, could solve the food shortages, raise living conditions above the crisis level, and provide the opportunity to address concomitant problems in transportation and other services. Resuscitating production, as we have seen, could not be achieved on central initiative alone, but local workers and institutions often proved unwilling to postpone immediate needs. In the textile industry, it was the national leadership of the union that would seize the initiative and begin to vie for the voice of authority. Its proposed remedy lay in establishing its own direction of activity in the industry while raising local personal and political consciousness through both practical experience and union cultural–educational work.[38]

Labor Discipline and the Workers' Courts

In the early weeks of Soviet rule, the dominant preoccupation of local actors with redressing past inequities had led only to a further breakdown of discipline. When the Petrograd Council of Factory Committees, for example, attempted to demobilize the wartime economy of the city in November 1917 without any program more concrete than a general commitment to workers' control, they fostered only additional disruption.[39] By April 1918, the All-Russian Council of Trade Unions considered the national situation serious enough to pass a resolution on "the complete disorganization of the enterprises and the absence of any labor discipline."[40] Developments in textile-producing centers reflected this pattern. In March, the Workers' Group of Centro-Textile declared the flax industry on the verge of closing "as a result of the general displacement produced by spontaneous demobilization and the beginning of the civil war."[41] By June 2, factory committee representatives from the textile enterprises of Moscow *Oblast'* concluded that "one of the primary reasons for the fall of labor productivity, aside from reasons of a technical and economic character . . . is the total disorganization within the enterprises and the absence of any kind of organizational and labor discipline."[42] Thus, after the *Sovnarkom* decree of June 28, "the [textile] factory committees, which received with the nationalization of the factories the right to supervise production and the right to sell cloth, sold the output of their factories without any sort of control, taking into account only their own factory and, in optimum cases, regional interests."[43] Consequently, when *Tekstil'shchik*, the official union journal, took stock of the accomplishments of the first year of Soviet power, it bemoaned the fact that working-class organs within the industry habitually competed among themselves, but could offer no corrective more specific than to endorse both additional central planning *and* greater freedom of action at the local level.[44] Oversimplifying matters but slightly, one observer attributed the poor state of labor discipline to a total lack of "comradely solidarity and discipline."[45]

Reports reaching the center from local factory committees and control commissions bore out these assessments and stressed that even the basic rudiments of discipline had eroded. At the Moscow Trekhgornaia Mill, workers habitually left early and in extreme cases extended their lunch break to as long as five hours.[46] Factory committee and union activists in Tver found it necessary to expend energy instructing textile workers on such mundane points as forbidding them to congregate on stairs before the end of work and admonishing them not to begin washing their hands more than five minutes before meals and the end of the workday.[47] Instructions issued in Moscow *Oblast'* spelled out even such elementary aspects of industrial life as the need to remain at work all day and to present an excuse in the event of absence, and it was deemed necessary to order workers specifically to refrain from drunkenness and card playing on the job and from mistreatment of machinery.[48] Although high-ranking proponents of centralization and decentralization continued to debate their positions, as the First *VSNKh* Congress at the end of May,[49] neither centralized nor decentralized organization seemed possible at lower levels. Local textile areas reported that their union "section," in reality, consisted of a few overworked individuals performing all of the tasks themselves; that commissions "died" as quickly as they formed;[50] that no union yet existed; or that a union had formed but established no ties to central bodies.[51]

Local officialdom, in these circumstances, was an eclectic mix. Experienced activists of working-class origin, inexperienced workers, office personnel, managers, and owners could all be found in its ranks. Some with no prior experience filled important positions. Others, who had held administrative positions before the revolution, found themselves promoted to responsibilities beyond their prior expertise or training. The incomplete data they furnished and the variations in local circumstances make a precise statistical description of their composition impossible, nor can one disaggregate officialdom by experience and prior employment. Yet, as this study shows, all these categories were represented, no single one was politically monolithic, and indirect evidence shows employees in particular to be widely divergent in their reactions to the revolution. Efim Gimpel'son has conducted the most thorough analysis to date on this question. In findings focused at the level of *glavki*, he finds more than half of the component in the textile industry to be "former workers," a term he does not refine further, and evaluates the participation of specialists as not high.[52]

Out of the concerns just enumerated emerged a number of attempts to raise labor discipline in order to increase productivity, and by midyear discipline became a major focus in union publications. Local sections such as Naro–Fominsk and Bogovsk began to organize *uezd* conferences to discuss the issue,[53] while national union spokesmen repeated that breaches of discipline harmed the working class as a whole. General suggestions in the union press included the implementation of Taylorism, productivity norms, and piecework wages.[54] On a more immediate level, this concern also led to a direct effort to make members of the local unions and factory committees the principal agents for restoring labor discipline: the formation of workers' courts. Soviet scholars credit Kostroma workers with originating the idea of comradely courts in September 1917, although by 1918 the unions and to a lesser degree the factory committees had become the engine of this program. On June 2, the Moscow *Oblast'* Conference of Factory Committees voted to establish workers' courts in each *raion* as well as a Norm Bureau to set pay scales for different types of work.[55]

Workers' courts carried the responsibility of redressing cases of indiscipline by workers as well as protecting the working class from abuses by the factory committees, foremen, and other supervisors. A general workers' meeting in each textile enterprise would, by open or secret ballot, elect three to five members to the courts for a six-month term, subject to confirmation by the *raion* administration of the union.[56] The courts met publicly and outside work time when possible, and the accused could call witnesses from the factory labor force. Such tribunals could issue warnings, give public reprimands, and authorize demotions and transfers. For flagrant or repeat violators, the courts could order transfers or suspensions, dismissals, permanent exclusion from production work, or even arrest and confinement for a period of one month. By the end of

July, according to one Soviet account, such courts functioned in most large textile factories and had also begun to adjudicate conflicts between factory committees and management.[57]

These efforts could be considered a limited success under the circumstances and, more to the point, were consistent with the union's commitment to reestablishing authority and discipline. When the textile apprentices of Tver organized an unauthorized strike in August, therefore, the Union of Textile Workers of Moscow *Oblast'* strongly denounced them. On August 11, the union specifically cited such "breaches of proletarian trade union discipline" as its primary motivation for strongly supporting the workers' courts.[58] The textile workers' union was, in fact, building credibility as a defender of discipline so rapidly that when *Narodnoe khoziaistvo* reported the Tver strike it made sure to stress the generally good record of the Union of Textile Workers and to treat the incident in Tver as an exception.[59]

Of greater significance, however, is the fact that the establishment of comradely courts demonstrated the depth of the differences between local organs and the rank and file. While we lack direct evidence on the composition of the courts, indirect evidence suggests that their membership derived from the same group that served on other local institutions. Given this, the courts seem, on the one hand, generally to have held with the factory committees against management. On the other, in the functioning of the courts the factory committees and local union sections acted more to restrain their respective constituencies than to represent them. As we shall see, this did more to raise animosity toward local officials than it did to restore productivity. Over time, the rank and file grew to view the courts much as they did the factory committees and local unions—more as a bureaucratic entity than as one representing them.

The Union Leadership and the Pursuit of Authority

The deterioration of factory productivity and local labor discipline in large measure underlay the determination of the national union leaders to establish their hegemony in the regulation of the industry. Formal divisions of authority notwithstanding, the union exerted itself simultaneously as the champion of working-class interests against the industrialists and as the engine of order and productivity within the industry itself. As Centro-Cloth increasingly lost public confidence and union activists prepared for the First All-Russian Congress of the Union of Textile Workers—agreed upon in December 1917 but convened only at the end of January—representatives of the textile workers continued to agitate in various venues for a larger voice in existing organs. On January 2, 1918, the Moscow *Oblast'* Delegate Conference of Textile Workers suggested the nationalization of the factories as an antidote to owners' sabotage. Their carefully worded resolution made it clear that the nationalization envisaged would occur only after an operative regulatory apparatus was put in place. Otherwise, the delegates warned, the step was "doomed to failure . . . [in view of] the extraordinarily low level [of development] of the laboring masses, the great insufficiency of skilled workers, and the total disorganization of the organs of regulation and supply. . . ."[60] When *VSNKh* reformed its Cotton Supply Committee on January 12 so that workers constituted two-thirds of its membership, the Workers' Group of Centro-Cloth seized the precedent and appealed to the Ministry of Food to recognize the *VSNKh* formula for all organs involved in regulating the supply of cloth. On January 18, the managing organs of Centro-Cloth, the Workers' Group, the Moscow *Oblast'* Committee on Cooperation, and the Moscow City Food Committee reached a preliminary agreement to distribute two-thirds of a new fifteen-member governing council of Centro-Cloth to the representatives of the workers.[61]

The mood emerging from national working-class institutions only encouraged such assertiveness. The Sixth All-Russian Conference of Factory Committees, held January 22–27 in Petrograd, evinced general support for more resolute action against ownership beginning at the level of the individual enterprise. At one extreme, A. Kakhtyn', a member of the Central Council of Factory Committees, argued:

> We cannot speak of *kontrol'* only in the literal, narrow meaning of the word; we must imply something broader and necessarily deeper; this is the regulation and organization of the whole economy, of all industry, of all production.[62]

> For us workers' control is not *kontrol'* in the direct meaning of the word, for us workers' control is something more. It is the active intervention [*vmeshatel'stvo*], it is management [*rasporiaditel'stvo*], it is the principle of the working-class administration of all industry, [and] together with the peasants of our whole economic life.[63]

Other speakers, however, did not share the confidence of Kakhtyn' that the workers were prepared for such steps. Indeed, the conference noted that when circumstances demanded "the actual implementation of *kontrol'*, we saw that some factory committees and trade unions were not able to do this, and the active workers did not have the physical possibility [of accomplishing this on their own]."[64] The majority therefore agreed that owners' resistance to the revolution gave the workers no choice except to assume direction of the national economy themselves,[65] but—and this was the critical point—only after creating the requisite apparatus of regulation.[66] Hence, on January 22 the conference paradoxically resolved "to carry out the nationalization of all enterprises simultaneously and immediately . . . [for which] the working class must create an organization whose goal is to prepare a planned and immediate nationalization of industry."[67] In the process, the factory committees coined a more fluid definition of nationalization that fell somewhere between the connotation of working-class supervision in the 1917 guidelines of state capitalism and the full confiscation enacted subsequently by the nationalization decree of June 28.

Thus, when the First All-Russian Congress of the Union of Textile Workers (January 29–February 2, 1918, O.S.) opened, it did so in a climate favorable to extending its mandate to reorganize Centro-Cloth[68] into a more thoroughgoing transformation of the industry. This obviously also bolstered the union predisposition to expand its authority in the industry. Thus, in his keynote speech on the morning of January 29, Rudzutak broadened the purpose of the congress by directing delegates "to speak *on all questions* of the regulation of the textile industry in conjunction with the transformation of the entire national economy."[69] G. M. Mel'nichanskii, who represented the Moscow Council of Trade Unions, supported him by emphasizing that organizing production now overshadowed even such traditional union goals of better pay and hours.[70] Ultimately, A. S. Kiselev of Ivanovo-Voznesensk, a veteran of the

Bolshevik underground who chaired the Centro-Cloth Workers' Group, translated these general prescriptions into a specific call for a completely new regulatory organ. Kiselev attacked both the credibility and the effectiveness of Centro-Cloth. He summarized workers' general complaints and enumerated the organ's specific failures in coordinating distribution. It received but 60 percent of the registered output, he noted, thus abetting rather than curtailing speculation. Moreover, he argued, such registered output came from only 220 of the more than 1,000 Russian textile enterprises, a further measure of the organ's ineffectiveness. Finally, Kiselev added that Centro-Cloth dealt only in cotton cloth, while separate manufacturers' committees still existed for other spheres of textile production. No voice defended Centro-Cloth, and its demise became an accomplished fact.[71]

Before it could create a new regulatory organ, however, the congress had to resolve for its own industry the issue of factory committee versus trade union hegemony just addressed by the First Congress of Trade Unions and the Sixth Conference of Factory Committees. This question, in fact, commanded equal attention with the reorganization of Centro-Cloth. Mel'nichanskii had included a proposal to combine factory committee and trade union work in his opening remarks.[72] Kiselev implied a need for greater union responsibility when he accused the factory committee conference of advocating the rapid nationalization of industry prematurely.[73] The key moment arrived at the evening session of January 30, when a general discussion of the definition of *kontrol'* escalated into a full debate over the authority of the factory committees. Proposals ranged, at one extreme, from creating a single, centralized organ of control under union direction to driving those referred to as the bourgeoisie from the factories and turning all functions over to the workers, at the other.[74] On the following day, the delegate Gladyshev complicated the picture further by expressing local reservations toward *any* direction from above. He declared frankly that the textile factory committees placed little faith in any central regulatory institution and felt a much closer affinity to the local unions.[75] Nevertheless, the decisions already made at the All-Russian trade union and factory committee meetings largely preordained the outcome of this debate at the congress, if not in life itself. The factory committees of the textile industry lacked both the material resources[76] and the personnel to resist formal subordination to the union.[77] The textile congress, therefore, declared the factory committees to be the basic unit of the union, gave higher union organs authority over factory committee elections, and required the committees to submit records and protocols of their meetings to the union.[78] Although the debate had made it clear that not every factory committee would accept such a transformation passively, the decision cleared the way for the organization of a new regulatory organ, Centro-Textile.

Thus, the Congress of the Union of Textile Workers exploited the prevailing climate to displace Centro-Cloth, but we should also note that key leadership figures expressed serious reservations about too hasty an implementation of nationalization and workers' management. When the union thus created Centro-Textile, subject to approval by *VSNKh*, the manner of Centro-Textile's formation addressed both attitudes. The congress issued "Regulations" (*Polozhenie*) that, in addition to subordinating Centro-Textile directly to *VSNKh*, entrusted it with full responsibility for the textile industry and empowered it to supersede all existing regulatory organs (Centro-Wool, Centro-Yarn, etc.) presently still outside the jurisdiction of Centro-Cloth. Centro-Textile was also to create its own organs for each branch of textile production as well as bodies to purchase raw materials, account for and distribute materials and semifinished goods, and confiscate individual textile plants where necessary. The "Regulations" further directed Centro-Textile to establish regional affiliates, Raion-Textiles, that would work with the regional councils of the national economy. Within Centro-Textile itself, the "Regulations" divided authority between a plenum, which the Workers' Group dominated numerically, and a presidium, in which owners' representatives held nine of eleven positions. Finally, and most important for affirming additional union authority within the industry, the "Regulations" stipulated that the Workers' Group in Centro-Textile and those in the Raion-Textiles follow explicitly the directives of the Union of Textile Workers.[79]

Centro-Textile began to function immediately, at least on paper. On February 2 (O.S.),[80] the union convened a special conference to define more clearly the new spheres of authority in the factories. Already empowered to elect three to five-member control commissions, the factory committees immediately received the authority to disperse troublesome commissions and to elect new ones where necessary.[81] On the same day, the Workers' Group of Centro-Textile met under the chairmanship of A. S. Bubnov, a member of the Bolshevik Central Committee, to elect its provisional presidium. As early as its second meeting, held February 18 (N.S.), the presidium of the Workers' Group began to discuss the creation of a state monopoly on cloth, and on February 26 it forbade the removal of goods from any factory without the permission of Centro-Textile. On March 1, a joint meeting of the Workers' Group of Centro-Textile and the presidium of the Union of Textile Workers began to plan the formation of Centro-Textile's regional organs.[82] Later that month, the *VSNKh* presidium formally approved the "Regulations" issued by the union congress[83] and on April 1 issued its own *Polozhenie*. The *VSNKh* version affirmed the union's formation of Centro-Textile and, in addition to previous stipulations, reserved for the workers two-thirds of the votes on any question in Centro-Textile, regardless of the proportion of their attendance at any given meeting.[84]

Centro-Textile did not conduct all of its business on paper, however, and its early activities began to foreshadow weaknesses that would later develop into larger problems. First, including workers' and owners' representatives in the same organ did not ensure that they would cooperate. On the contrary, Centro-Textile rapidly became an arena in which the two groups replicated the confrontations over management prerogatives that were taking place within virtually every textile factory at the time. First, Rudzutak complained that the former industrialists' committees (Centro-Yarn, Centro-Wool, Centro-Flax, and others) resisted inclusion in the new institution and that the remnants of these organizations systematically opposed the Workers' Group in Centro-Textile once incorporated.[85] Second, the fact that the workers' representatives outnumbered the owners only aided the latter in crystallizing into an organized opposition within Centro-Textile.[86] Third, its large parliamentary structure turned out to be too unwieldy for people inexperienced in such bodies. The Workers' Group, consisting of representatives of diverse working-class organizations, had difficulty reaching a consensus. In July, the union replaced this thirty-seven-member body, criticized as resembling a "bourgeois parliament," with a smaller group of seven.[87]

In practice, Centro-Textile more commonly followed national trends than set them. Some projects, it is true, began promisingly enough. By the end of March 1918, for example, Centro-Textile had conducted preliminary inspections of 346 cotton enterprises in Moscow, Vladimir, Kostroma, Riazan, Tver, and Iaroslavl provinces,[88] but even such limited successes were rare. In fact, the All-Russian Council of the Union of Textile Workers found it necessary as early as April to reaffirm the authority of Centro-Textile in expropriated factories against the encroachments of the union's own local sections by declaring that no union organization had the right to establish a separate organ to rival Centro-Textile's authority.[89] At the same time, enterprises as well as individuals resorted to expedient measures. In January 1918, some factories responded to the contracted supply of cotton by sending out their own delegations to acquire materials by any means available, including black-market purchases.[90] In March–April 1918, the Centro-Textile Trade and Distribution Department bartered cloth directly for bread from the Commissariat of Food, and if in need of money Centro-Textile sold cloth on the private market.[91] In a rare show of unity, the presidia of the Union of Textile Workers and Centro-Textile passed a joint resolution warning of the danger of tolerating such a large proportion of textile goods entering the black market,[92] but such measures contained more style than substance. By mid-November 1918, Tseitlin of the All-Russian Council of the union would report to that body that state-owned flax factories still purchased raw materials from private individuals instead of relying on the state distribution network.[93] Obviously, such practices only exacerbated the problem of speculation.

In the spring of 1918, owners' recalcitrance moved to a new level as well. From the October Revolution until February 1918, the country underwent the Red Guard Attack on Capital, a period dominated by direct expropriation at the factory level with virtually no central direction. Given the workers' propensity to take revenge for past grievances, this was a time of uncertain personal safety for owners and managers and, consequently, a time of flight for large numbers of them and their families. As a result, the proprietors and technical personnel hostile to the regime who remained learned to replace futile confrontation with less direct forms of resistance. For example, managers of the Glukhov and Trekhgornaia textile mills in Moscow, citing the authority over production decisions reserved for them by the Decree on Workers' Control, declared their equipment exhausted and reduced operating speeds by 20 percent. Simultaneously, one of the industrialists' representatives in Centro-Textile introduced a resolution before the plenum to make a similar reduction mandatory for the whole industry. Workers' control, however, also empowered factory control commissions to establish production minimums, and since machine-operating rates directly affect output the workers' representatives in the mills overrode the decision. At the same time, a united workers' front in Centro-Textile defeated the proposal for a national reduction.[94] Prokhorov, proprietor of the Trekhgornaia, utilized a different stratagem: payment of wages in kind. Popular with many workers as a hedge against their reduced purchasing power, the practice had an obvious negative effect on the state regulation of distribution. In addition, by placing textiles in so many different hands, it also became easier for Prokhorov to slip goods to the black market through proxies.[95] Centro-Textile and *VSNKh* countered with constant reiterations that goods be surrendered only to authorized organs of distribution,[96] while on May 18 the Textile Section of the Council of the National Economy of the Northern Industrial Region felt compelled to limit the amount of finished goods that owners could distribute to twenty arshins per worker.[97] Meanwhile, in Ivanovo–Voznesensk the political unreliability of technical and office personnel as a group, the so-called employees, became so troublesome by April that the union gave up trying to enlist them into membership.[98] Problems, in short, outweighed successes.

Despite its dominion over the Centro-Textile Workers' Group, therefore, the All-Russian Council of the Union of Textile Workers rapidly lost confidence in Centro-Textile and by early summer began a public campaign to discredit it. During the week of June 20, several union leaders utilized a plenary session of the council to level criticisms. Rudzutak excoriated Centro-Textile for leaving the technical apparatus of the industry in the hands of local union branches, which, in his view, prevented the coordinated accounting of production. Rykunov attacked it for excessive bureaucratism, and Asatkin claimed that Centro-Textile still did not know how many factories, workers, looms, and spindles fell under its jurisdiction. Employing an approach that stressed regional rather than leadership concerns, the delegate Petrov, who represented the seventy-nine textile factories of the Northern Industrial Region, criticized Centro-Textile for working only for Moscow and doing nothing for other areas.[99]

With such criticisms mounting, Centro-Textile certainly did not need the additional obligations that the nationaliziation decree placed upon it one week later. The supply and management of all cotton plants with a basic capital of more than 1 million rubles, the wool, flax, silk, and jute enterprises with a capital of 500,000 rubles, and the hemp factories whose capital exceeded 250,000 rubles—about 80 percent of all concerns in the industry—became its responsibility literally overnight.[100] Even though an appendix to the decree ordered managerial and technical personnel to remain on their jobs under pain of trial before a revolutionary tribunal,[101] the transition was abrupt, disruptive, and ultimately beyond the capacities of Centro-Textile.

The union therefore seized the opportunity to become more assertive following the *Sovnarkom* decree. It immediately assembled the factory committees of the textile enterprises of Moscow *Oblast'* on June 28 to work out the implementation of the nationalization decree, even though the gathering produced little that was new. The meeting instructed factory committees and control commissions to prevent the removal of factory property, to deliver output only to Centro-Textile, and to file monthly reports. These directives quite obviously only repeated orders ostensibly already in force, and additional communications illustrated how little information the center had actually gathered about the various spheres of the industry during more than a half year of workers' control. The conference decided, for example, that workers' control should apply to warehouses as well as factories, that each factory committee or control commission would submit an inventory of factory property and supplies to Centro-Textile and the relevant Raion-Textile by July 29, and that local sections of the Union of Textile Workers would supervise the fulfillment of these directives.[102] On July 1, nationalized factories in Ivanovo–Voznesensk still had to be told exactly what data superordinate union bodies required. Not only were they to account for machinery on hand, as might be expected in anticipation of nationalization, but they were asked whether a cultural–educational section, socialist club, or committee of the poor had been established as well as whether even the most basic organs—factory committee, control commission, management—had organized.[103] On July 12, the Moscow *Oblast'* Council of the National Economy published instructions on technical and bookkeeping functions in the enterprises, knowledge that in theory factory organs had been utilizing since November 1917.[104]

In the second half of 1918, therefore, the Union of Textile Workers began to supplement its attacks on Centro-Textile with a bid to establish direct ties to local organs. As Tseitlin summarized matters on July 27,

> Today it needs to be said once and for all that the victory of the proletariat is in our hands. And those who have gathered here know well that if the textile industry cannot carry out its part the revolution will perish. . . . The textile industry must be restored and strengthened, and this cannot be accomplished by generating paperwork [*razsylki bumag*] but through vigorous interaction with the local areas. It is essential that our professional organ be in close contact with the organs which are being created [there].[105]

Indeed, the union would continually push itself forward as the institution responsible for the textile industry not only against Centro-Textile but also at the expense of the *VSNKh* centers and *glavki* by the end of 1918.[106]

Throughout the remainder of the summer of 1918, the additional steps taken were better suited to the preparation of a nationalization decree than the implementation of one already in effect, and they further illustrate how little attention was being paid central directives. While *VSNKh* declared the cloth trade a government monopoly on July 22,[107] three days later—and nearly a month after the nationalization decree—the Council of the Union of Textile Workers heard a report that owners were still transferring their factories to German and English ownership.[108] A union directive of August 1 essentially repeated the terms of the nationalization decree for the factory committees and warned yet again against carrying out a nationalization without higher authorization or in a manner that might disrupt production.[109] One week later, the Ivanovo–Voznesensk Union of Textile Workers reported that it was still carrying out preparations for full nationalization,[110] and in September Centro-Textile had not yet established the accounting and supervision over textile warehouses, a project that had been a high priority since June 28.[111] When *VSNKh* formulated a new plan for its subordinate organs in October, the document included a number of measures designed to centralize the purchase of supplies and materials "for the most speedy implementation of the decree of June 28. . . ."[112]

While thus exercising little authority at the local level, the Council of the Union of Textile Workers intensified its campaign against Centro-Textile. On July 25, Rudzutak charged that Centro-Textile tried to carry out too many tasks—technical work, finance, supply, regulation—by dealing directly with factories instead of using a regular apparatus. After speaking in favor of a greater devolution of responsibility to the Raion-Textiles, he scorned Centro-Textile for its failure to create "anything positive in the sense of a plan for its future work."[113] On the following day, the council heard the complaints directly from the representatives of the Raion-Textiles themselves. The delegate Timofeev of Ivanovo–Kineshma Raion-Textile criticized Centro-Textile for its reluctance to delegate any but the most menial tasks. He claimed, with more than a little bravado, that his own institution had firm plans for the distribution of textile goods that were not implemented only because of Centro-Textile's jealous monopoly of authority. Speaking for Moscow Raion-Textile, the delegate Matveev explained that no clear division of authority existed between the central and *raion* levels of work, a situation that Centro-Textile used as a rationale for ignoring the *raion* organs altogether.[114]

The August issue of *Tekstil'shchik* featured an article that advanced the union attempt both to wrest full authority in the industry from Centro-Textile and to redefine the union's purpose. The author, identified as Mel'nik, argued in favor of extending the scope of economic and administrative activities against those who still supported retaining the union as a so-called fighting organization, that is, a partisan organization working solely on the workers' behalf. Echoing Rudzutak's criticisms of Centro-Textile, he denounced its preoccupation with trivial tasks better handled by local organs and chastised it for the lack of production, accounting, and distribution plans. Knowledge of how to finance factories was lacking, he added, while speculation in textile goods flourished in Moscow and Petrograd as well as in remote areas. He proposed greater participation in administration from the *raion* organs but pointed out that the failure of Centro-Textile to complete the creation of a full apparatus made even this impossible on a national scale at present.[115] Thus, by late summer the union had heard several calls for the dissolution and replacement of Centro-Textile,[116] and in November V. P. Nogin, a member of the Bolshevik Central Committee and future director of the nationalized textile industry, essentially declared this to be the official policy of the union.[117]

Centro-Textile also experienced encroachments on its prerogatives from above when *VSNKh* began to assume greater responsibility for the nationalization effort in August–September 1918. During these months, *VSNKh* directed factory committees and control commissions in carrying out more detailed prenationalization inventories that would provide the information Centro-Textile had been instructed to collect in February. More specific than the cotton plant inspections Centro-Textile had conducted the previous spring, these inventories for *VSNKh* provided data on the type of factory, capital, available supplies, location, size of work force, composition of the management, and in many cases commentary on the workers' attitude toward nationalization.[118] Although Centro-Textile had produced a few conspicuous successes, as when its Flax–Hemp–Jute Department participated in every step of the nationalization of the important Demidov Mill in September,[119] it accomplished but a small fraction of its required task. As the first anniversary of the revolution approached, Centro-Textile and the union leadership had neither established regular administrative relations with subordinate levels nor achieved harmonious cooperation among themselves.

This gave rise to a reexamination of the union from within. In May, A. Davydov, reiterating an essentially Menshevik position, argued in *Tekstil'shchik* that unbridled growth, caused by the present practice of enrolling members simply by withholding dues from their wages, weakened the union. He held that the Union of Textile Workers would improve qualitatively if membership were again made voluntary.[120] M. Aleksandrov countered in the next issue, however, that Davydov's fears about encroachments on the workers were unfounded. The union, he held, would inevitably become a state organ, since it was a working-class organ within a workers' state. Davydov, in the opinion of Aleksandrov, should worry not about what a trade union is but about whom it serves.[121] This conflict not only extended existing differences about the role of the unions after the revolution but, more to the point, anticipated future battles. In 1920–1921, the Workers' Opposition would raise the issue of the proper function of the unions and the status of the union leadership relative to national, especially party, institutions. With such differences unresolved in the ranks in 1918, the union appeared from the local perspective to move in several directions simultaneously. At a time when the All-Russian Council of the Union of Textile Workers spoke boldly of taking full charge of the industry,[122] the union leadership could also be found passing resolutions urging local sections to maintain internal discipline, as in refraining from initiating strikes.[123]

The Revolution between Center and Factory

Developments in the regional institutions, local organs, and textile factories followed a self-defined logic. The workers and their representatives had anticipated a rapid improvement in material conditions after the revolution through a more equitable distribution of goods and services. As *Tekstil'nyi rabochii* made clear early in 1918, the union expected a revitalization of exchange between town and countryside in the near future as well as the elimination of money wages. Workers' control, characterized as a "transitional stage toward socialism," would eradicate practices such as the hoarding of cloth by proprietors that kept prices high.[124] Demobilizing the wartime economy, the conventional wisdom of early 1918 ran, would help to make industry more responsive to consumer needs and liquidate unemployment.[125]

Such expectations gave rise to local initiatives that complicated the process of defining the locus of authority. At one extreme, *Tekstil'nyi rabochii* found itself obliged to address seriously the arguments of those who advocated solutions grounded simply in basing new regulatory institutions on existing prerevolutionary committees and commissions.[126] At the other, the workers who took over factories on their own, especially in enterprises where peasant-workers constituted a large proportion of the work force, frequently treated the plants only in a proprietary manner.[127] Thus, after October 1917 actors who had previously internalized no strong ethic of professionalism or self-discipline were suddenly presented with the opportunity to redress grievances on their own. They largely saw no contradiction in their simultaneous attempts to enact punitive measures against the former owners and managers and to pressure central authorities to accelerate the long-term aims of the revolution.

Local activists had largely shared with the Bolshevik leadership the assumption that the transfer of economic authority would be uncomplicated, but proposals for the immediate future differed dramatically. Kakhtyn', reflecting one view, had told the Sixth All-Russian Conference of Factory Committees in January that "it is necessary to create the highest organs *from above*. And at the bottom the organs need do nothing other than transmit information, various data, figures, and so forth."[128] Other socialists shared his optimism on the ease of the task but by no means endorsed the letter of his proposals. We have already noted the Left Communists' reservations to centralizing proclivities, which the Bolsheviks' Left Socialist-Revolutionary collaborators largely shared. Mensheviks and SRs outside the ruling coalition continued to spread their own viewpoints as well, and anarchists still enjoyed access to broad public forums in early 1918. Indeed, in rebuttal to Kakhtyn', the anarchist Bleikhman asserted that "any work, individual or mass, must proceed from bottom to top and not top to bottom."[129] In a speech interrupted several times by applause, he declared that

> the idea of nationalization has nothing in common with the workers. Get this into your heads [*zapomnite*] once and for all. The idea of nationalization is clearly a counterrevolutionary bourgeois idea; it will be carried out in order to enslave the workers further, to enslave further the toiling peasantry—this is what the idea of nationalization will create for us. . . . The Bolsheviks have become the chameleon of the bourgeoisie; they promote nationalization instead of socialization.[130]

Support for a revolutionary transformation thus proliferated; it was more difficult to achieve unanimity of purpose.

Practical obstacles also exacerbated this lack of consensus. It proved especially difficult to recruit specialists who were not irreconcilably hostile to the new order. The idea of using existing expertise in the service of the revolution was not limited to Lenin and his faction in the Bolshevik leadership. Local support for the idea, in the form of recurrent resolutions of factory committee conferences in the Moscow region in November–December 1917, specified that employees and technicians be included in workers' control, and the Executive Committee of the Moscow Soviet stipulated that engineers, technicians, and accounting personnel be included in factory control commissions.[131] In January 1918, the Commissariat of Education began to employ specialists not only for the transformation of the economy but also to create an apparatus for long-term planning.[132] At present, however, far too few technical specialists enlisted to satisfy the existing demand, and worker–specialist conflict outstripped cooperation.[133]

Revolutionary institutions did not respond with any single strategy. Technically, *VSNKh* bore the responsibility to comat bourgeois sabotage through its subordinate *sovnarkhozy*, but its sluggish pace of organization and assertion of authority made this impossible in the immediate future. Friction between the owners and workers had reached serious proportions long before *Sovnarkom* created *VSNKh*, and local councils of the national economy formed too slowly to effect the situation significantly thereafter.[134] In the Moscow textile factories, therefore, local union sections and factory control commissions attempted to fill this void. By February 1918, control commissions had formed in 73 percent of Moscow's enterprises, but they could battle lockouts and other obstructions only on a factory-by-factory basis,[135] a condition characteristic of the rest of the Central Industrial Region as well.

Consequently, despite the increasingly greater assertiveness of the national leadership of the Union of Textile Workers throughout the initial year of Soviet rule, its All-Russian Council did not speak for a cohesive institution. The union had registered 571,400 members nationally by mid-January,[136] but growth did not translate into disciplined action. Indeed, reports generated at lower levels retained a strong strain of localism. When the Ivanovo-Kineshma Union of Textile Workers met on January 25, wages rather than organizational questions dominated its agenda.[137] In

Tver province, union organization outside the city of Tver had not even begun as late as February 1.[138] Nor did activists display any inclination simply to do the bidding of the central authorities. In addition to the frictions between the Ivanovo–Kineshma and national officers already examined, the factory committee representatives of Moscow also showed themselves preoccupied with immediate concerns when they met November 28, 1917: pay and achieving authority over production for the *raion* soviet.[139] As Lozovskii, a member of the All-Russian Council, reported to the national union congress at the end of January, the hastily and "spontaneously" organized workers were not yet closely tied emotionally to the union or well integrated into its structure. At the beginning of February 1918, in his view, the most important organizational work still lay ahead.[140]

In the textile industry, the multiplicity of local aspirations clearly affected this process of organization. Like Lozovskii, activists at all levels impatiently called for a higher degree of organization, but they did not always act in a manner that would bring their words to fruition. At a time when the union leadership was advocating the unification of authority, its own increasing criticism of Centro-Textile made the further formation of Raion-Textiles and provincial organs, Gub-Textiles, something of a futile gesture. We have already seen that the representatives of the Raion-Textiles held little except resentment and contempt for Centro-Textile, and they interpreted organizational mandates from above in light of their own inclinations. Also, the few Raion-Textiles in existence refused to work harmoniously with the councils of the national economy, as directed in the *Polozhenie*. In fact, the Council of the Union of Textile Workers reported in April that the Raion-Textiles had declared themselves subordinate only in local matters.[141] In one instance in May, a union section created a Raion-Textile not as an extension of the authority Centro-Textile but as what they declared to be an alternative to the general frustration of dealing with the bureaucratized central institutions in Moscow.[142]

Such independence only invited the wrath of the union leadership. In July, I. I. Korotkov, a Bolshevik veteran from Ivanovo-Voznesensk, argued before the All-Russian Council that nationalization was too important to entrust to the Raion-Textiles and that they should confine themselves to preparatory work, a sentiment

Asatkin seconded. On the same occasion, the union official Frolov cautioned against devolving any responsibility whatsoever on the Raion-Textiles and proposed that the union itself create new administrative groups (*kusty*).[143] Union opposition did not deter Centro-Textile from creating additional bodies, however, and by the end of the year nineteen Raion-Textiles operated, albeit in a stormy fashion. In one case, a *raion* union so strongly resisted the creation of a new Raion-Textile that only four votes prevented its defeat, despite the personal intervention of a Centro-Textile representative on behalf of the new organ.[144] Throughout the fall, local factory committees complained so frequently about the redundant nature of the Raion-Textiles' activities that Centro-Textile decided near the end of 1918 to create new Raion-Textiles only as departments of local councils of the national economy.[145] As the influence of Centro-Textile declined in the second half of the year, the already precarious effectiveness of the Raion-Textiles diminished further.

　　Moreover, the redefinition of the scope of union activity that Davydov and Aleksandrov argued in *Tekstil'shchik* did not significantly affect the task of local organization. The union in Moscow *Oblast'* still found it necessary to dispatch field representatives widely to try to incorporate local unions into the larger union apparatus,[146] and a rise in membership brought with it an increased number of complaints about the separatist tendencies of local union activists.[147] Local sections defiantly communicated that the workers' confidence in the union would continue to decline unless the general economic situation improved.[148] Prospects for achieving this, however, grew less likely when owners' and managers' intractability increased after June 28. Indeed, centers and *glavki* found it necessary to dismiss professional specialists whose loyalty they questioned in spite of the urgent need for trained personnel, and at the end of July the All-Russian Conference of Textile Workers passed a resolution calling for the purge of former industrialists from the Centro-Textile apparatus.[149] *Sovnarkhozy* even at the provincial level complained that qualified staff were hard to find,[150] and reliable cadres were so scarce in the textile factories that in some cases illiterates served on factory committees, even in the city of Moscow.[151]

　　The textile industry competed poorly for those who had expertise to market. As shown by the industrial census of 1918, specialists gravitated toward the urban areas and the most developed spheres of industry. To be specific, 10 percent of the work force concentrated itself in heavily industrialized Petrograd. At the same time, 12 percent of the available employees located there, as did 26 percent of those classified by the census as highly qualified specialists and 34 percent of people with a higher education. Conditions in Ivanovo–Voznesensk, the leading textile-producing city, contrasted sharply. There one found 12 percent of the workers of the province but only 7 percent of the employees, 4 percent of highly qualified specialists, and 3.5 percent of those who had completed a higher education. To employ a different comparison, metalworking and machine construction employed 20 percent of the workers of the nation and 24 percent of its highly qualified specialists. The textile industry attracted about 40 percent of the workers at the time of the census but only 31 percent of the available specialists and production managers.[152]

The View from Below

The conditions discussed thus far reduced the ability of working-class institutions to influence attitudes in a desired direction even at the local level. Shortages of personnel and resources prevented the union from conducting the very cultural–educational programs on which it counted to raise the workers' consciousness. One union instructor, after an inspection tour of several provinces in December 1917–January 1918, foreshadowed a trend when he described the educational programs of union organizations then being formed as "weak."[153] In February 1918, when the union assembled a commission to evaluate educational work, the overwhelming majority of representatives of local union sections agreed that present efforts were failing. The commission held that the participation of the "intelligentsia" was absolutely vital to success but that such people participated only rarely. Moreover, the commission found, the workers had not understood the few attempts that had been carried out. If cultural–educational work were to succeed in the local areas, it would have to be "popularized."[154] These assessments, brought forward by local union officials, were obviously shaped by their firsthand experience in attempting to address the rank and file with existing strategies. The national union, therefore, adapted its educational efforts to all workers, not only those with socialist sympathies, with a new program that emphasized approaching the workers through recreational outlets. If clubs were established, the argument ran, where workers could drink tea, play cards and chess, make music, and relax, they might also take advantage of reading materials in the club library. Lectures, when given, should be limited to one hour, and tactics in general should be modified to recognize that textile workers as a group were not accustomed to being organized.[155]

After midyear, nevertheless, union cultural–educational work stood still. In October 1918, the plan that the union's Cultural–Educational Department presented to the All-Russian Council focused above all on finding enough instructors so that work in the provinces could actually begin.[156] From a regional cultural–educational conference in Simbirsk province, the rapporteur Moshkov communicated that cultural–educational departments did not exist below the *raion* level as of November 1918. Despite priorities that included organizing adult literacy classes and workers' clubs, his department had by this date only circulated a letter to factory committees—which elicited but scant response—urging them to select members for this work. At the same conference, the delegate Shelenshkevich complained of the workers' general passivity and criticized those who refused to work for the local soviet, apparently a large proportion, for a lack of consciousness.[157]

Enlightening the workers, however, stood little chance of success—in popularized form or otherwise—unless material conditions improved. Instead, the situation worsened. World War I had stretched the workday to fourteen and fifteen hours as women, children, and invalids were pressed into factory work in greater numbers.[158] When production fell sharply in 1917, however, many lost these jobs, and by one count 100,000 unemployed resided in the city of Moscow alone in 1918.[159] Wages continued to suffer as well. In September 1918, the textile worker of Moscow *Oblast'*, if still employed, earned less than counterparts in other industries. While the monthly average in metal fabricating stood at 374.57 rubles, workers in the various spheres of textile manufacturing survived on considerably less: silk, 329.94; cotton, 264.08; wool, 305.64; flax–hemp–jute, 256.11.[160]

Indicative of the general dilemma was the position of the Moscow City Soviet. Virtually powerless to cope with unemployment and hunger, it tried to address one chronic problem it considered within its capability to affect: poor housing for workers. In November 1917, the Moscow Soviet nationalized all large-scale rental dwellings as a prelude to the complete nationalization of housing the following year. This action redistributed housing without improving conditions as the newly confiscated dwellings fell rapidly into disrepair.[161] In November 1918, *Pravda* reported that in the Krasnaia Presnia textile district of Moscow an estimated 10,000–13,000 inhabitants lived in but forty-two large dwellings nationalized earlier in the year.[162] Another account in the textile union press affirmed that in Moscow, which the author considered the center of the workers' movement, remedial housing measures produced little result. As before the revolution, four adult textile workers and five to eight children crowded into rooms eight by twenty feet, and as many as 150 workers occupied cockroach-infested foul-smelling dormitories. The real source of irritation—and this point deserves special stress—was that neither the factory committees nor higher union organs responded to rank-and-file entreaties to alleviate such hardships. The author quoted workers who attested that complaining about conditions led to expulsion from the factory, which prompted one mother of five to ask rhetorically where she might then turn.[163] As this example illustrates, the factory committees were resistant and even defiant to higher authority, but they also ran roughshod over their own constituencies when pressured.

Early attempts at social legislation produced similarly poor results. As cotton shortages in early 1918 dramatically increased unemployment,[164] the hard-pressed textile workers would be among the leading beneficiaries of measures to retire invalids at half pay and to create workers' sanitaria.[165] In May–June, proposals to limit the workday once again to eight hours and to curtail the hiring of children directly addressed abuses that had reappeared in the industry during World War I.[166] The Union of Textile Workers tried to ease unemployment by transferring surplus workers to areas where they might be needed[167] or to perform alternative work at full pay.[168] In practice, few of these measures advanced beyond the stage of good intentions.

This state of affairs left factory committee, control commission, and local union representatives caught between two levels of pressure. We have already noted the local suspicion of central direction expressed at the first national union congress, the defiance of national union leaders by local union sections in the matter of running expropriated factories, the tendency of local organs to ignore communications and their refusal to furnish replies, the recurrent problem of locally initiated strikes, the routine dominance of wage and other material issues over union matters in the reports filed by local sections, redundant factory committee complaints about the Raion-Textiles, and, in general, the suspicion with which local organs greeted overtures from the center. Local institutions nevertheless sought material aid, moral support, and instruction from above, although superordinate bodies could furnish little. From below, the officials of these working-class institutions encountered increasingly vehement criticism. Lacking an alternative acceptable to them, they continued to strive to maintain independence before superordinates but increasingly appeared before the local constituency as a new and frequently unresponsive officialdom.

Left almost completely to their own devices, therefore, mass workers and local activists turned to speculation, theft, and the use of force to meet their needs. In the city of Moscow, the state tolerated open profiteering at the notorious Sukharevka Market out of an inability to provide a workable alternative, and on Grigoriev Street in Ivanovo individuals hawked scarce foods equally brazenly.[169] As much as two-thirds of the food supply and half of the nonfood products eluded state attempts to control distribution.[170] On November 28, 1917, the Ivanovo–Voznesensk Soviet appointed commissars to protect supplies and prevent the removal of output from factories,[171] and on December 2 the Ivanovo–Kineshma Soviet instructed railroads to ship goods to no one other than specified soviet boards in Ivanovo–Kineshma *raion* and Vladimir province.[172] While they took such measures ostensibly against the propertied classes, the need to establish strict control over access to the keys of all factories, warehouses, and offices in one region of Vladimir province, for example, indicates that theft and speculation served a broader constituency than just the owners.[173]

The disappearance of supplies of food and raw materials brought on predictable responses. In the Ivanovo–Kineshma region, where there were virtually no manufactured goods to trade for grain in 1918, peasants refused to honor the deteriorating ruble. By late spring and early summer, therefore, textile workers there grew more receptive to the idea of seizing grain directly, especially after the state introduced ration cards in April, and detachments for this purpose were well into operation by fall.[174] On June 30, a conference of Moscow *Oblast'* textile factory committees resolved not only to honor the traditional practice of closing plants for a summer break but would schedule the two-week hiatus to coincide with the period of peak agricultural work. In addition, workers who went to the countryside could, with the approval of their respective factory management, remain an additional two weeks at half pay.[175] In spite of such palliatives, general anxiety increased. When representatives of the Commissariat of Labor met with the textile factory committees of Tver on August 8, the reports from the individual factories practically without exception spoke only of food shortages.[176]

The additional strain brought on by the outbreak of full-scale civil war reversed those minimal gains previously achieved. In Kineshma *raion*, factory control commissions had created an atypically efficient distribution network in early 1918, and on May 20 a conference of control commission representatives founded a *raion* body to account strictly for supplies passing between *raion* and provincial centers. Almost immediately, it successfully distributed 280 wagons of raw cotton received by barge from Nizhnii-Novgorod. When the war broke out, however, the commission reported that it felt its future sources of supply, and therefore its effectiveness, were in jeopardy.[177] Shortages of fuel and materials indeed became a recurrent theme in *Tekstil'shchik* during the summer of 1918,[178] and the loss of Turkestan to the Whites halted deliveries from what had been the principal domestic source of cotton. By autumn, one frequently expressed opinion held that only a concentration of supplies could slow the pace of further factory closings.[179] Aleksei Rykov, chairman of *VSNKh*, convened a special plenum in September to discuss emergency distribution measures to combat closings.[180] The task contained a conundrum. Concentrating supplies in a few capable factories made sense from the perspective of conserving scarce resources, but it actually exacerbated the problem of establishing authority over individual factories. As Kiselev reported to the September *VSNKh* plenum, mills acquired considerable leverage once they received the few resources available, and they used the threat of closing themselves to resist subordination to policies from above.[181]

By the end of the year, specific complaints about the plight of women in the industry supplemented the more general reproaches on the condition of the industry. In *Tekstil'shchik*, an advocate of better treatment for women in the industry, Mikhailova, suggested that an explanation of the plight of females lay in the fact that "women constituted a small percentage of the members of the union."[182] The union, Mikhailova charged, was more concerned with politics than with the welfare of women in the industry, and in her view women deserved more attention. Female textile workers had, she noted, begun to replace legal marriages with common-law marriages even before any state pronouncement on the subject. This step, which her tone inferred was a manifestation of instinctive enlightenment, produced mixed results. It removed the woman from the moral and economic domination of a male head of household, but it also left women to support children alone. Moreover, it broke what Mikhailova considered the natural tie between the woman and the home. Shouldering sole responsibility for herself and her children condemned the female textile worker to a life of drudgery under present conditions and precluded any raising of consciousness. She was more likely to be concerned with questions such as finding an alternative to leaving young children alone during working hours or in the care of another child than with politics. As a nonnegotiable first step, Mikhailova advocated the establishment of nurseries and elementary schools.[183]

Indeed, the status of women as much as any single issue combined immediate material concerns and the long-term considerations of transforming society voiced by the revolutionary intelligentsia. The 1,000 delegates who attended the First All-Russian Congress of Women Workers on November 16 called for new living spaces to be constructed outside the city. They also resolved to ban work for children under age sixteen and to shorten the workday for youths over sixteen. Since such provisions were supposedly already in effect, one can only conclude that they were not being observed in individual factories. Unlike many other gatherings of the period, however, this conference also directed attention to social questions of a more permanent nature. For example, if prostitution were deeply rooted in capitalism, the conference resolved, it was not enough to close brothels and prosecute procurers. Such women needed both material aid and new social forms, that is, "the replacement of the bourgeois family by free marriage."[184]

Conclusion

Viewed from the perspective of central institutions in 1918, the Soviet state found itself torn between utilizing measures to establish its authority, on the one hand, and attempts to restrain local excesses and premature initiatives, on the other. In the textile industry, central officials remained convinced that the long-term goals of building socialism—the nationalization of industry and the institution of workers' management, in particular—should proceed only after the necessary administrative apparatus was in place. When the threat of external interference accelerated this timetable in June, however, the state turned increasingly toward a reliance on bureaucratic solutions to the task of organizing the economy. While dependence on the bureaucracy might not have caused the Bolsheviks ideological concern before the revolution, the institutions that emerged in practice functioned differently than anticipated. Centro-Textile specifically failed to enlist or coerce the assistance of the owners and managers in sufficient numbers, to establish a strong presence of workers' representatives in the regulation of the industry, or to extend the authority of *VSNKh* over this branch of industry. By the first anniversary of the October Revolution, the national leadership of the Union of Textile Workers had taken firm steps to assume authority in the industry, but the effective exercise of such was as yet in the planning stages.

The shortfalls of central administration were ultimately rooted in local conditions. Prerevolutionary expectations that local workers would participate actively in the creation of a revolutionary order were called into serious question in the first year of Soviet rule. Local priorities eluded any central control during the Red Guard Attack on Capital, and matters did not change significantly thereafter. As communications from local activists continually reiterated, constituents at the factory level concentrated on their own concerns: finding enough to eat and seeking a better place to live. In pursuit of this, however, there appeared an additional cleavage between the rank and file and the perceived indifference of even local officials. Thus, as central organs in the textile industry moved toward a greater exercise of decision making in 1918, local areas moved in the opposite direction. The behavior of local textile workers indicated that their representatives were empowered to communicate their grievances but raised serious doubts about the degree of direction officials held over local affairs.

4

The Revolutionary Experiment Matures: Centralization and Localism

The first anniversary of the October Revolution witnessed the end of one period of revolutionary history and the beginning of a second, comparatively more focused stage. The initial year of precarious power had given rise to varying degrees of revolutionary exhuberance and excesses. Impatience and miscalculations at all levels had colored the trial and error that characterized the Red Guard Attack on Capital, the midyear crisis, and the overtures toward the state direction of production and distribution that followed the *Sovnarkom* decisions of the spring and early summer of 1918. The period from late 1918 to mid-1919, by contrast, brought forward attempts to assimilate the lessons of these false starts and to convert them into workable programs and institutions. Ongoing crises, ideology, and the demands posed by the civil war continued to play an important role, of course, but a new consideration also appeared. This was the attempt to synthesize prerevolutionary expectations with the actual experience just gained during a year of trying to consolidate the revolution.

Thus, as the anniversary of Red October passed, national institutions of the revolutionary state furthered policies that, in their degree and manner of success and failure, would do much to shape the Soviet future for the next decade. Disappointed at the low degree of cooperation with initiatives from above, central institutions responded in October 1918–July 1919 by relying even more on the propensity to aggregate authority at the top of state administration and less on the state capitalist tenet of devolving responsibility on lower organs. This resulted in a highly situational definition of centralization. On the one hand, national institutions concentrated decision making in their own hands, a step that intermediate and local organs were virtually powerless to prevent. On the other hand, superordinate bodies lacked anything even approximating adequate means to implement, by suasion or coercion, policies and decisions thus formulated.

This version of centralization was particularly pronounced in the Russian textile industry. In this sphere, national policies and local aspirations continued to clash regularly as immediate material concerns obscured the longer view. Indeed, as the second revolutionary year began, the sway of mass workers in the industry assumed an even greater role. Not only did the daily pressure they exerted individually and collectively on local officials dominate communications and relations with higher bodies, but mass workers themselves entered administrative posts in significant proportions. Differences of awareness and attitude continued to separate the conscious activists from the rank and file on the level of political consciousness, but in the work of routine administration distinctions between mass and conscious workers became less discrete since employees, experienced activists, and workers with no prerevolutionary experience all found themselves pressed into filling local offices, often inadequately. In the process, main attention continued to fall on the same issues highlighted during the winter of 1917–1918: reviving production, through nationalization if necessary, and solving the food and raw materials supply problems by whatever available means.

Centralization Ascendant

By mid-1919, the military and economic crisis reached its most severe level. The Bolsheviks withstood the initial threat of a rapid White victory but paid a high price for their own battlefield successes thereafter. Approximately half of all party members served in the military, making up between 5 and 10 percent of the Red Army.[1] The Soviet Republic in 1919 was reduced approximately to the size of sixteenth-century Muscovy. Forces under Kolchak and then Denikin mounted major offensives during the year that strained the full resources of the Soviet Republic. Political and military control of the borderlands proved elusive, and anti-Bolshevik cossack governments ensconced themselves in the Don, Kuban, and Terek areas of the south and southeast. Ural and Orenberg cossacks joined forces militarily, and former members of the Constituent Assembly established a government at Samara.

The war complicated other problems. Food and industrial supplies continued to contract, additional factories closed, and the escalation of the exodus of the urban population reduced Moscow and Ivanovo-Voznesensk to a fraction of their former size. Qualified cadres became ever more scarce in the textile-manufacturing centers, as the most talented left either for the front, to find better markets for their skills, or simply in search of food. In the countryside, meanwhile, the repeated military reversals, conscriptions of peasants by both Reds and Whites, and the formal beginning of forced grain requisitions—decreed by *Sovnarkom* on January 11, 1919—destroyed what little order remained.

This national emergency strengthened the position of the advocates of centralization in the upper reaches of economic administration and the unions, and the state decision to concentrate on the organization of industrial trusts at the end of 1918 moved this dimension of the revolution into a new stage. It was neither that past organizational successes made possible further progress toward state capitalism nor that the civil war made centralization necessary but, as *Narodnoe khoziaistvo* explained matters, because a transition "from decentralization to centralization" provided an alternative to "the lack of organization in the center and the spontaneity in the local areas."[2] Indeed, the Second All-Russian Congress of Councils of the National Economy in December 1918 declared this to be the most pressing task of the moment[3] and liquidated its *oblast'* councils made "superfluous" by "the general strengthening of the central apparatus of economic power."[4] In mid-January 1919, the Second All-Russian Congress of Trade Unions eliminated its *oblast'* organizations in the same manner.[5] This did not, however, achieve the desired relations between the center and the lower levels. Instead, the result was "glavkism." This was the name given the extreme centralization of industrial decision making that struck strong root throughout the *VSNKh* apparatus in first half of 1919.[6] Indeed, the greater centralization of decision making became dominant in all areas of political and economic administration at the time.

The Eighth Party Congress, held March 18–23, 1919, also reflected this propensity toward greater centralization. Its economic program, the first formal revision since 1903, reflected the greater emphasis on central direction already being attempted in practice and singled out the unions for greater organizational responsibility.[7] The congress also moved to make the direction of party cadres more systematic. Iakov Sverdlov, who had previously directed party personnel affairs with only a small staff, died of Spanish influenza shortly before the congress opened, and replacing him occasioned a revision of procedures. The Central Committee took over party personnel assignments and announced its intention to liberate party organizations from dependence on the soviets.[8] At the same time, the party undertook its first administrative purge by launching a reregistration of members, a step that produced a short-term decline in party rolls.[9] By December 1919, however, the campaign also reversed one of the party's more glaring weaknesses by establishing regular contact between the Central Committee and 95 percent of its district organizations.[10] Thus, the party itself assumed a more assertive posture in questions of organization, although not until 1920 would it be able to put forward a significant campaign to establish party organizations in enterprises that lacked them, even in major urban areas.[11]

These were the conditions in which the union leadership attempted to assume chief responsibility for the textile industry, an aspiration that entailed maximizing the authority of the union heads without necessarily enhancing the power of top state and party authorities. The critical turning point occurred late in November 1918, when the union assault on Centro-Textile already in progress led to the establishment of a second regulatory institution in the industry, Glav-Textile. In theory, this new organ was a department of Centro-Textile responsible only for the administration of nationalized plants. In practice, it took instruction directly and exclusively from the union leaders. Consequently, from the perspective of the union leadership Glav-Textile possessed one significant advantage over Centro-Textile: the industrialists had no direct voice within it. Thus, when Glav-Textile assumed accountability for the industry's nationalized textile enterprises on December 2,[12] the industry began to experience a new variation on dual power: Centro-Textile, a Central Administration (*Glavnoe upravlenie*) under *VSNKh*; Glav-Textile, a Central Management (*Glavnoe pravlenie*) only technically subordinate to Centro-Textile and, to repeat the point, responsible just for nationalized enterprises. This precarious hierarchy, however, proved temporary. Glav-Textile quickly increased the number of nationalized plants under its jurisdiction, thereby enlarging its sphere of influence and, by extension, enhancing the power of the union. Glav-Textile was able to do this rapidly because, as noted earlier, the number of officially registered nationalizations was far lower than the total of enterprises actually expropriated. Therefore, if central institutions had officially registered no more than four textile nationalizations in any single month through the end of October 1918, the pace significantly accelerated thereafter: November, 14; December, 39; January 1919, 38; February, 72; March, 71; April, 64; May, 27; June, 79; and July, 24.[13] The formation of Glav-Textile in this way transferred greater responsibility for the industry to an organ under exclusive union direction, in contrast to the system of sharing authority with the owners in Centro-Textile.

As a result of its early operations, Glav-Textile earned a reputation as one of the most effective centralizing institutions of Soviet Russia,[14] but not simply through the wholesale nationalization of factories. As it brought greater numbers of factories under its authority, it also created *kusty* (production groups) to unify enterprises under a single collegial administration. In actual fact, widespread nationalization and the organization of production groups occurred almost simultaneously and as parts of the same process. Of the forty-eight *kusty* that existed in the textile industry by April 1920, Glav-Textile created forty-seven in the first half of 1919.[15] In the textile industry, therefore, the increased centralization of decision making created, at least formally, a national framework for administration by July 1919 where none previously existed.

The key problems of the industry, to be sure, transcended decision making. The Whites held cotton-producing Turkestan for most of 1918–1919. This obviously exacerbated the already serious national supply problem. Throughout 1918, nevertheless, rumors of impending foreign relief, including the importation of cotton from America, circulated. In the spring of 1919, union rapporteurs squelched these rumors and delivered nakedly frank appraisals of the "catastrophic" supply situation in all aspects of textile produc-

tion.[16] While national industrial productivity fell to 44 percent of
1913 levels in 1918 and decreased again to 21.6 percent in 1919,[17]
the textile industry contracted even more severely. In Moscow
province, textile production in 1918 fell to 24.6 percent of prewar
levels and to 12.7 percent in 1919. There remained in 1919 only 68.8
percent as many textile workers there as in 1913.[18]

   In the local areas and factories, self-reliance prevailed. This produced some striking examples of the resilience of the human spirit
in the face of adversity but much more frequently provided a
chronicle of despair and friction. Talented officials were in short
supply, and those who filled local posts frequently did so out of self-
interest rather than revolutionary or civic zeal. Hunger and illness
led increasing numbers to flee to the countryside, and those who
remained at the textile plants endured conditions that communications regularly described as worse than those in tsarist times. Moreover, the former owners and managers proved resilient in adapting
to conditions. Although reliable and detailed composite data on
their representation in the new apparatus are unavailable and were
almost certainly never compiled, constant complaints from the
factories and other indirect evidence attest to a continuity in management and administration resented by the rank and file. In point
of fact,

> A "white" professor who reached Omsk in the autumn of 1919 from
> Moscow reported that "at the head of many centres and *glavki* sit
> former employees and responsible officials and managers of busi-
> nesses, and the unprepared visitor to these centres and *glavki* who
> is personally acquainted with the former commercial and industrial
> world would be surprised to see the former owners of big leather
> factories sitting in Glavkhozh, big manufacturers in the central
> textile organization, etc."[19]

The Demise of Centro-Textile

The textile union, to be sure, did not accept these conditions
passively. Its opposition to Centro-Textile helped produce a partial
redefinition of authority relationships within the industry in late
1918 and foreshadowed future battles inside and outside the party.
To be specific, the campaign against Centro-Textile provided the

union leaders the opportunity to attempt to increase their influence vis-à-vis *Sovnarkom* and the Bolshevik Central Committee as well as in relation to lower union organs and rival institutions. In the process—and these issues would receive a national hearing in 1920–1921 in the intraparty conflict over the Workers' Opposition—the union officials also began to assume actual authority over the nationalization of industry from *VSNKh*, although the appearance of formal subordination to the Supreme Council of the National Economy would continue to be observed.

What transpired graphically demonstrates the key tenet of administrative relations in this period: that institutional assertiveness far surpassed formal authority in importance. On one level, the union leaders greatly reduced the influence of industrialists and their representatives on top-level decision making in the industry by making Glav-Textile functional. As we have seen, owners by no means disappeared from the middle and lower administrative apparatus altogether, since the absence of alternative managerial expertise made their complete removal impracticable. Nevertheless, in 1919 the owners lacked the power to prevent the union from curtailing their weight at the highest level of the industry. On a second level, the union usurped functions to which *VSNKh* and the soviets also laid claim, again encountering largely verbal resistance. As Glav-Textile gained authority, however, *VSNKh* increasingly ratified its actions. Finally, union initiatives ran counter to the spirit and letter of directives on administrative cooperation issued by the national leadership. As union efforts built a record of relative success, however, national officials tolerated them. As already noted, it was at this time that Lenin, Rudzutak, and others publicly lauded textile administration as a model for other industries.

By autumn 1918, the union campaign against Centro-Textile reached a peak. A pivotal moment arrived on October 17, 1918, when the All-Russian Council of the Union of Textile Workers, acting apparently only on its own authority, voted to merge Centro-Textile with the *VSNKh* Department of Fibrous Substances, ostensibly to facilitate the implementation of the nationalization decree. On the surface, the main prerogatives of Centro-Textile did not change. It retained the initiative for the administration of the textile industry, the distribution of finished goods, the financing of enterprises, and supply. The degree of impatience behind the council's action, however, was evident in the fact that the official union account also repeated the long-standing complaint that establishing local organs of Centro-Textile still lay in the future and reiterated an already familiar proposal to combine factories for the sake of efficiency.[20]

One month later, the union leaders abandoned indirect language altogether. In meetings held November 13–14, the council voted to create the Central Management, Glav-Textile, which the union hoped would answer directly to *VSNKh* for all nationalized textile factories.[21] In nationalized plants, this new organ would assume Centro-Textile's responsibilities for supply, finance, organization, and regulation, and, if the union leaders had their way, supersede Centro-Textile at the local level by creating *kust* production groups. As outlined at these meetings, each *kust* would combine factories in close geographic proximity that would perform every function of the production cycle, from raw-material procurement to final output distribution. An administrative collegium of five to nine members, of whom technical personnel were to constitute at least one-third, was to head each production group. The factory committees of member plants would elect one-third of the group administration, the provincial *sovnarkhoz* a second third, and the *oblast'* council of the national economy the remainder. Group administrations, in theory, selected the three to five member managements of individual factories, subject to confirmation by the local union.[22]

In the immediate future, only part of this plan would succeed. On November 16, the *VSNKh* presidium confirmed the union leadership's creation of Glav-Textile and appointed to it the seven figures nominated in the union resolution: Rudzutak, Nogin, Korolev, Rykunov, N. I. Lebedev, A. A. Ganshin, and N. M. Matveev.[23] However, the fact that *VSNKh* failed to subordinate the organ directly to itself, as recommended by the union, caused immediate consternation among the Glav-Textile leadership. Because of this *VSNKh* decision, only part of the plan to supersede Centro-Textile had so far succeeded. When Nogin reported this to Glav-Textile's first formal meeting on November 20, Lebedev refused to accept the action of *VSNKh* as final, and after further discussion the group decided to confront Centro-Textile on this matter directly.[24] On December 1, Glav-Textile used the occasion of its initial public declaration to press its case further. "To All the Workers of the

Textile Industry" announced that Glav-Textile intended nothing short of assuming full direction of the industry. It laid claim to the prerogatives of Centro-Textile and added that it would institute a general economic plan uniting all large-scale, functioning textile enterprises.[25] *Tekstil'shchik* endorsed this in late December by calling on the forthcoming Second Congress of the Union of Textile Workers to designate "a single regulatory organ" for the industry and by enumerating in detail Centro-Textile's failed attempts to establish links to the local areas.[26]

Glav-Textile challenged Centro-Textile functionally for the first time on January 1, 1919, by trying to take responsibility for financing nationalized factories. Since it had never established the orderly pattern of accounting on which efficient finances could rest, Centro-Textile was vulnerable to this assault. A Glav-Textile report, in fact, reiterated the well-established charge that Centro-Textile did not even know what goods were on hand. It consequently allowed itself to be cheated by issuing credit not only on the basis of goods being manufactured at the time but for prerevolutionary inventories as well. In numerous cases, Centro-Textile advanced funds to individual plants on this basis, only to have deserting owners abscond with them. Even after Centro-Textile created a Control Department to curb such abuses, Glav-Textile added, financing became no more businesslike. Enterprises found it necessary to dispatch delegations directly to Centro-Textile to attain funds, and the most assertive and worldly factory activists petitioned on an almost perpetual basis, to the detriment of plant operations.[27]

A stormy and confused joint meeting of Glav-Textile and the union leadership on January 22 produced the next major attack.[28] Nogin, chairman of Glav-Textile, reopened the offensive by complaining that Centro-Textile treated Glav-Textile as just another of its departments and refused to relinquish control of finances. For its own part, Nogin added, Glav-Textile had earned greater authority and independence by virtue of its success in creating production groups during its very first weeks of existence. The principal problem, in his view, could be traced to the *VSNKh* decision to make Glav-Textile a subsidiary of Centro-Textile. Other speakers reacted in a sympathetic but disjointed manner. Glav-Textile members Rykunov and Lebedev agreed that the central task was to liberate Glav-Textile from Centro-Textile but differed in their proposed methods. Asatkin and I. I. Kutuzov of the All-Russian Council, on the other hand, apparently were unaware even that *VSNKh* had altered the initial union proposal of November 13–14, and both made statements that Centro-Textile had already been disbanded. Under these circumstances, the meeting found it necessary to pass two resolutions: that Glav-Textile was the single organ governing the industry and that Centro-Textile should be liquidated without delay.[29]

On this basis, the All-Russian Council argued assertively for centralizing textile production and distribution under union direction at the Second All-Russian Congress of the Union of Textile Workers but with incomplete success. The topic of centralized organization certainly entered every discussion of the congress, held January 27–February 1, 1919. Nogin, once again articulating the resentment against the continued influence of the prerevolutionary industrialists, complained that more than a half year after the nationalization decree the textile plants still remained in the hands of their former owners and managers "and in reality did not come under state jurisdiction." Rudzutak numbered driving out of the spirit of private ownership among the immediate priorities of the union,[30] and Rykunov, disregarding consistency of metaphor, characterized Centro-Textile as "new wine . . . in an old apparatus."[31]

The central issue, however, turned out to be more than simply one more denunciation of Centro-Textile, and the delegates[32] as a group would not support without reservation the centralizing propensities of the council. The congress did not agree to disband Centro-Textile and rejected the union council's proposal to concentrate authority in Glav-Textile. It decided instead to create yet another new organ—the Central Committee of the Textile Industry—to oversee a number of fairly autonomous departments in the textile industry.[33] Ideally, this new institution restructured Centro-Textile without specifically involving Glav-Textile and thus arousing further the opposition at the union congress to expanding the authority of the union leaders. Also, since the *VSNKh* presidium and the union leadership jointly appointed the responsible officials, subject to final confirmation by *VSNKh*, the new arrangement would resolve the confusion over the November 13–14 union resolutions without reopening that troublesome discussion as well. Finally, the Central Committee of the Textile Industry would receive all of the Centro-Textile prerogatives that the union leaders had previously tried to transfer to Glav-Textile, while Glav-Textile retained the right to authorize nationalizations.[34] In short, the creation of this committee gave the appearance of a concession to the opponents of centralization.

Rather than hindering the centralizers, however, the creation of the Central Committee of the Textile Industry produced the opposite effect. In actual fact, the new committee never functioned as designed, the Glav-Textile disregarded its nominal independence. References to it in existing documents are so frequently contradictory that it is clear that officials within the industry did not fully understand its status and apparently confused it with the Central Committee of the Union of Textile Workers. What is most important is that *VSNKh* specified that "the majority of the members of the Committee [of the Textile Industry] must be part of the *Glavnoe pravlenie*,"[35] which undercut any attempt to limit central union authority. Of the original eleven appointees to the Central Committee of the Textile Industry, five—Rudzutak, Nogin, Lebedev, Matveev, and Rykunov—were Glav-Textile members, with the possibility that a sixth would soon be added.[36] In addition, the union veteran Movshovich received an appointment as well.[37] Thus, although the local delegates who opposed centralization had been momentarily appeased at the second union congress, in the aftermath the proponents of central authority in Glav-Textile and the union leadership acted as if they had received a mandate rather than a defeat.[38]

As early as mid-March, these supporters of concentrated central direction began to win their point as Glav-Textile assumed sole authority in the industry. Although *VSNKh* would not formally acknowledge the hegemony of Glav-Textile in the industry until October 1919,[39] the new institution began immediately to absorb competing regulatory organs. In a candid communication to a local factory in mid-February, Glav-Textile outlined its intention to unify all plants in production groups and "to plan the work in the factories, coordinate it with the general economic plan of a communist society and to make production organized."[40] Departments of Glav-Textile began taking over parallel organs of the Centro-Textile apparatus in March, a process that continued throughout the year. It pursued this without extensive interference since "the majority of the members of the [Central Committee of the Textile Industry] were also members of the Central Management; the [Central Committee of Textile Enterprises] existed almost nominally[;] the *Glavnoe pravlenie* spoke and acted in its name."[41]

Such administrative maneuverings did not in themselves resolve the persistent questions on the agenda, however, and within Glav-Textile organization lagged even at the highest levels. When Glav-Textile met on March 19, Lebedev observed that its internal division of responsibility "has hitherto had a disorganized character. There is no precise distribution of work among the separate members of the Management."[42] Other speakers supported him, and P. V. Grinevich of the Glav-Textile Administrative Office put forward a recommendation to form a Control-Organizational Department, an internal inspectorate reporting directly to the chairman of Glav-Textile.[43]

Despite these internal fissures, Glav-Textile displayed a confident posture in its public dealings. On March 21, Nogin seized the question of the traditional Easter break to make clear that Glav-Textile would not confront adversity in the passive manner of Centro-Textile. He sent a telegram to individual factories that challenged the widely held belief that shortages of fuel and raw materials dictated that the plants close as in previous years. Instead of accepting the situation, Nogin's telegram sought information on whether a general shutdown was widely desired and, if so, for how long; whether it was possible to curtail production in order to avoid closing; and whether Glav-Textile could transfer workers affected by closings to different tasks.[44] In late May, Matveev responded assertively to a report from the Administration of the Cotton Factories of Tver that mills still turned to the black market as their only source of materials. Concerned that competition among state factories for goods only increased black-market prices, Matveev provided a proposal as old as the revolution itself: creating a single purchasing organ.[45] Establishing such an organ in a single industry could not, of course, solve a problem of the magnitude of the national black market, but the move was more than symbolic. On June 2, Matveev, as chairman of the new purchasing organ, tightened controls—and provided insight into the full scope of abuses—by announcing that Glav-Textile would honor only requests for payment that included the address of the person to be paid.[46] At the same time, the *Glavnoe pravlenie* established fixed prices for finished textile products and attempted once again to take over the distribution of cotton goods.[47] These were, to be sure, steps that above all indicated the extent of prior misconduct and the proportions of the task ahead. They were also, however, actions that showed Glav-Textile was prepared to back up its grandiose statements with specific albeit modest measures.

Nothing illustrated this determination more than the establishment of over forty *kust* administrations by July 1919. Ostensibly founded to provide planning and unification in the industry, these production groups also attacked institutional parallelism. Since local areas possessed a plethora of competing organs, Glav-Textile began to try to extend its own authority by formulating a production plan for each *kust*. On June 7, Glav-Textile devoted its entire meeting to "The General Plan of Work of the Silk Industry,"[48] and on June 11 committed itself solely to studying the work plans of nine individual production groups.[49] It conducted similar sessions on June 19[50] and June 26.[51]

Nationalization and the Production Groups

Changes in the pattern of the nationalization of textile factories were linked closely to the growth of Glav-Textile. The new administrative organ could not have undertaken even its modest endeavors had it not significantly increased the number of nationalized factories under its authority. After October 1918, Glav-Textile greatly accelerated the registration of nationalized plants, and, even more significant from the perspective of central leaders, these transfers were now widely initiated "from above" rather than "from below."[52] This is not to say that the consolidation of enterprises into nationalized production groups proceeded as smoothly as Bolshevik rhetoric and some public resolutions of the period would suggest. It did signify, however, that the lessons learned from the general displacement that preceded and followed the nationalization decree caused officials to put greater emphasis on realistic coordination by the end of 1918. Some officials, such as *VSNKh* Chairman Rykov, still regularly spoke as if the state had systematically incorporated whole spheres of industry into a centralized administration in the first six months after the nationalization decree.[53] Such hyperbole, however, began to share attention with more sober assessments, such as the call from the textile factory committees of Iakhromo (Moscow province) for "a transition from nominal to actual nationalization" in December 1918[54] and the open observation that despite nationalization a significant number of Moscow factories remained in private hands.[55] A recurrent sentiment at the Second Congress of the Union of Textile Workers was, in fact, that early optimism about the ease of nationalization had been unfounded.[56]

By late 1918, therefore, the nationalization process itself began to change in important ways. *VSNKh* records show that by early November the textile industry considered each nationalization separately,[57] with mixed results. The process still focused on individual enterprises rather than state monopolies, but at least *VSNKh* now authorized a nationalization before the event instead of ratifying the accomplished fact, as it had in early 1918. Under these conditions, factory committees or control commissions that completed a prenationalization inventory and questionnaire could petition to nationalize a plant, and such requests became common.[58] By December, however, the volume of work so increased that *VSNKh* began to nationalize a series of factories in a single decree.[59] In part, this resulted from an inability to scrutinize each petition, some of which undoubtedly sought only official sanction for expropriations carried out earlier in the year. One cannot conclude from this, however, that the process always lacked order.[60]

The larger role that Glav-Textile and *VSNKh* officials played in the formation of factory managements at the moment of nationalization illustrates a measurable increase in their actual authority. In some instances, creating a management entailed overriding local desires to exclude remaining prerevolutionary owners and experts from administration or sending union instructors to explain procedures and supervise proceedings.[61] When necessary, *VSNKh* would remind factory managements of their responsibilities within the plants as well as their obligation to cooperate with superordinate institutions.[62] Beyond this, in mid-December 1918 the Union of Textile Workers addressed the ubiquitous tension between factory committees and managements by issuing a new general instruction explaining in detail the division of authority. The management would direct finance, supply, personnel assignments, and technical-administrative functions, while the factory committee completely supervised work rules and pay norms, labor discipline, cultural-educational work, and questions traditionally within the competence of the union. Both shared responsibility for the workers' well-being. The union instruction also replaced the factory control commission with an inspection commission (*revizionnaia komissiia*) empowered to monitor all functions but not to overrule management.[63] Of greatest concern, and also a source of ongoing central-local friction, *VSNKh* worked to guarantee that persons with technical expertise occupied one-third of the management positions.[64] Finally, by the end of the year it became clear that Glav-Textile rather than *VSNKh* was further enlarging its role in supervising the formation of managements in nationalized factories, since it simply began to inform *VSNKh* of actions taken in this area by memorandum after the fact.[65]

In the first half of 1919, both nationalization and the formation of production groups moved forward definitively. In January and February, the *VSNKh* presidium extended the practice of nationalizing a series of textile factories in a single action,[66] authorizing, for example, the expropriation of nineteen Petrograd textile plants on February 11[67] and of forty enterprises at its meeting of March 13.[68] Typically, the inner leadership (*Malyi Prezidium*) of Glav-Textile considered a nationalization petition, the full *Glavnoe pravlenie* reviewed approved recommendations, and *VSNKh* provided the final confirmation,[69] sometimes accompanied by specific assignment to a production group.[70]

Actual nationalizations consequently began to resemble more closely the process as described in theory, yet the formation of production groups reflected the primacy of practical considerations. The idea of production groups was not itself new. The Council of the Union of Textile Workers had entertained proposals to form *kusty* as early as July 1918.[71] The initial Glav-Textile meeting of November 20 included a debate among those who favored forming production groups on local initiative alone (Lebedev), as a joint central and local endeavor (Nogin), or by sending instructors from the center as supervisors (Rykunov).[72] The actual formations, however, occurred only after Glav-Textile became fully operative in December. Guidelines issued January 5, 1919 stated that only Glav-

Textile and the Union of Textile Workers could organize production groups, that they performed the task as equals, that factory committees joined them in naming the group administration, and that the most immediate attention was being paid to areas where *kusty* were deemed most needed.[73] Temporary instructions issued by the union on January 27 made the group administrations' jurisdiction more specific: authority over all movable and immovable property, including living spaces; the right to distribute semifinished goods within the *kust* for further processing and to supervise finished goods; the responsibility to attain a general plan of work, which it could endorse or reject, from all firms in the group and to synthesize a group plan; the power to transfer materials and personnel within the group; and the obligation to submit minutes of all meetings to Glav-Textile.[74] Officials such as Asatkin, a member of Glav-Textile and of the council of the union, personally encouraged local officials to take over their Raion-Textile apparatus once a production group was formed,[75] a right to which the First Congress of Group Administrations of Textile Enterprises formally laid claim on March 6.[76]

Union Leaders Retrench

Quite obviously, such an organization of textile administration ultimately responsible to union authority coud be only as effective and disciplined as the union itself, and the Union of Textile Workers took decisive steps in November 1918–July 1919 to solidify itself internally as well. The rapporteur on the question of organization, Afanasev, admitted retrospectively at the second union congress that early operations had lacked coordination. Although a national union had technically emerged in January–February 1918, the four *oblast'* organizations that performed the majority of the work did not actively function until June, and then did not work compatibly.[77] Stability eluded even the top echelons. Membership changed regularly not only in the All-Russian Council but also in its executive organ, the Bureau of the All-Russian Council of the Union of Textile Workers. Every member of the bureau elected in January had departed for a more responsible position in government or *VSNKh* by May.[78]

Well into 1918–1919, therefore, the union leadership found itself repeating basic steps. The union established itself in virtually every factory and *raion* by the end of 1918 and recruited almost all workers and employees in the industry, but this still amounted to little more than the registration of personnel.[79] According to a union report filed January 8, 1919, a work load that forced organizers routinely to cover several different regions prevented them from completing any single project. They would carry out a basic organization, convene a regional conference, and move on without establishing links between the new sections and the broader union apparatus.[80] The redundant character of union communications testifies that these conditions led to a low level of compliance with directives. The important council meetings of November 13–14, 1918, for example, reiterated in detail once again the need for factory committees to obey government and union decrees.[81] In the final days before the Second Congress of the Union of Textile Workers, the leadership still found itself repeating that there could be but one all-encompassing industry union.[82]

The union congress that convened January 27, 1919 held far-reaching debates on the centralization of authority within the union, therefore, just as it had on general economic administration. At the opening session, the delegate Kovalevskii, who identified himself as a Social-Democrat Left Internationalist, reopened the issue of whether the union was an independent or state organ.[83] As we have seen, this was in no way a new issue. In contrast to previous debates, however, the ensuing discussion encompassed not only ideology and competing perceptions of proper union activity but also the implications of local union independence. Discussion of this issue by 1919 manifested a deepening not only of differences over central–local authority but also brought out the fact that the same union leaders who advocated concentrated authority within the union also championed flexibility for the union itself in relations with national state and party institutions. The majority at the congress agreed that the union could not yet carry out efficiently the tasks that status as a state institution implied and that it would do better to confine itself to organizing the masses and raising their consciousness. Indicative of the local ambivalence toward the center, however, the delegates did not endorse complete local independence, since they felt this would only leave lower organs vulnerable to the influence of the hostile class elements in the factories.[84] The congress reached no binding conclusion.

As in the area of economic regulation, the union leadership increased its authority once the congress adjourned. On February 7, the Central Committee of the Union of Textile Workers moved to strengthen its Moscow branch by consolidating its three separate sections into one.[85] The March conference of the Moscow *Oblast'* union voted to merge its apparatus with that of the national union.[86] On a national scale, the union leadership reduced the number of member branches (*otdeleniia*) from seventy to twenty-six.[87] This aimed at eliminating inefficient smaller sections, ensuring that the structure of provincial and *raion* unions replicated the structure and functions of the parent body, and dividing authority more systematically at the lower levels. At the time, local presidia and instructors still performed almost all union functions below the *raion* level. If successful, this reorganization would create a structure of departments even in local union sections, at least in large-scale factories.[88]

The Central Committee of the Union of Textile Workers could not pursue its objectives with full energy, however, because of internecine disputes. Absenteeism plagued the ten-member Central Committee elected February 1. As few as six discharged all responsibilities from the outset,[89] and union accounts three months later reported no improvement.[90] In early May, Rudzutak, Rykunov, Bychkov, Dement'ev, and Kisel'nikov walked out,[91] an act that left the remaining members divided as well.[92] Union records do not make clear whether absenteeism alone caused the breach—the official union account lists the disaffected as having left "without sufficient reason"[93]—but *Tekstil'shchik* singled it out for specific mention while alluding cryptically to additional "other causes."[94] Setbacks notwithstanding, the reconstituted leadership pushed forward in the summer of 1919 toward planned production, and on July 17 it optimistically commissioned representatives of its Labor and Organizational Departments to submit an industry plan by August 5.[95]

Problems at the lower administrative levels undermined the central union leadership as well in the first half of 1919. Leverage deteriorated as factory closings and defections caused membership to fall from more than 642,518 on January 1, 1919, to 484,747 on

July 1.[96] Of greater import, redundant reports continued to reach the center of organizational programs that never began in earnest or of isolated pockets of efficiency undermined by the chaos surrounding them.[97]

Failures of the *Kusty*

Glav-Textile and union officials, consequently, could not take for granted the loyalty of their constituency, either in provincial and lower organs or on the floor of the factory. By 1919, the All-Russian Council of the union openly admitted that provincial and *raion* sections had to become the real bearers of union policy if it were to succeed[98] but feared the implications of such a statement. The center, in fact, became keenly sensitive to whether the widespread local strikes and demonstrations occurring within the Soviet Republic were articulations of antigovernment sentiment or simply economic protests. Early in the year, for example, the union instructor A. P. Lomonosov hastened to reassure his superiors that a demonstration of textile workers in the town of Viazniky stemmed only from the absence of food and not from anti-Bolshevism.[99] The flour and other goods that the Moscow *Oblast'* council of the union rushed to Viazniky pacified the workers, but the reports of similar protests that followed immediately from Ivanovo–Voznesensk gave rise to new calls for a more comprehensive solution.[100] The Orekhovo–Zuevo Section of the union also reported that food strikes regularly took place in the factories under its jurisdiction. In May, the workers in one of the united Nikolskii Factories staged a particularly severe disorder, and when conditions did not improve by October they walked out again. Two months later, the symbolically important Likino Mill struck as well.[101] These were not isolated occurrences.

Viewed from the perspective of the center, therefore, local officials, in addition to the unpredictable rank and file, presented a considerable problem. The central textile institutions were dissatisfied with their local officials and—paradoxically—wanted greater numbers of them. We have already encountered the resistance to centralization manifested by local representatives at the Second Congress of Textile Workers. Even if local activists had complied more closely with the desires of the center, however, there were too few talented and qualified persons among them to complete even the minimum tasks on the scale of the whole industry. This was more than a repetition of the chronic tsarist complaint of a shortage of trained personnel. In provincial, regional, and local institutions, nothing resembling the full complement of owners, managers, technicians, and office staff who kept the industry in operation before the revolution remained by 1919. Those who had not fled Russia altogether could find more remunerative work in other spheres of industry. In many cases, the few remaining specialists shared authority in the textile industry with persons of little or no supervisory experience.[102] As the annual report for 1919 of the Glav-Textile Trade Department stated matters:

> Frequent mobilizations, diseases reinforced by cold and hunger, the flight of workers and employees to the provinces and their defections to other institutions [and] the introduction of piece-work pay in a disguised form have led to a reduction in the size and quality of the staff of Glav-Textile. The more or less experienced and knowledgeable staff who have left have been replaced by inexperienced ones, mainly women.[103]

Other Glav-Textile organs also lacked technicians, engineers, bookkeepers, and instructors,[104] and even *Tekstil'shchik* could not fill its staff.[105] Neophytes everywhere stepped into responsibilities that they often discharged in keeping with the only model they knew, the arrogant and proprietary manner of the tsarist bureaucrats.

In November 1918–July 1919, it became abundantly clear that the centralization of decision making would not alone provide the solution to the industry's disorganization. The leadership of the union and Glav-Textile had pushed both for the inclusion of technical personnel in administrative bodies and for greater authority at the center of the textile industry. In theory, these positions did not contradict each other. Lenin's perception of state capitalism had stressed the need to utilize prerevolutionary specialists until workers assimilated professional skills. In the textile industry, however, regulating appointments to *kust* and factory administrations proved to be far different than controlling the actions of the appointees. Technical personnel frequently undermined the wishes of superordinate organs and opposed the desires of the local workers as well. Thus, as state organs searched frantically for talented administrators, one by-product of the situation was that charges of sabotage became more impatient.

Industry leaders found themselves consistently on the defensive for retaining the very personnel from whom the local rank and file felt the revolution would bring them relief. In November 1918, the local delegate Smirnov reported with unmasked contempt to a meeting of the Romanov Flax Mill in Iaroslavl province that in Moscow former directors still found positions on factory managements and that they continued to oppose the workers' elected representatives.[106] Discussions at the second union congress repeated similar sentiments so strongly that Nogin took the floor on January 30 to defend yet again the use of professional and technical experts.[107] A large part of its problem stemmed from the fact that the center could not adequately screen these personnel politically. On February 25, the Central Committee of the Union of Textile Workers declared that retaining persons as employees in factories they previously owned was in principle unacceptable. This should occur only in extraordinary circumstances, each instance should be weighed on its individual merits, and in any event owners could remain only with the approval of the local section of the union.[108] The *VSNKh* journal *Ekonomicheskaia zhizn'* continued to feature articles that attempted to define the proper role of technical specialists,[109] while *Tekstil'shchik* complained that "elements alien to the working class" still participated in industrial administration.[110] When the union central committee met on July 8, it noted that individuals hostile to the working class received appointments to plant managements despite union directives that only those well known to the union could serve.[111]

The insufficient number of workers and employees willing to work for the revolution, in brief, simply did not perform as anticipated in revolutionary guidelines. The union continually complained of a shortage of instructors,[112] and local improvisations generally produced poor results. As one instructor reported from Tver province in January 1919:

> Matters stand worst of all in [the city of] Tver itself. . . . Today a so-
> called *Raion* Secretariat was formed and its reports display a

complete lack of understanding of its position and a total ignorance of what was created since one member says one thing, another says something else and evidently no one knows what to do. There are no ties to the group administrations, no directives were issued and each member acts on his own, not knowing satisfactory work from unsatisfactory work, as one of the Secretariat said.[113]

Superordinate offices, nevertheless, exacerbated the problem by raiding lower organs for talent. At a time when the Glav-Textile Wool Department complained of losing its qualified cadres to institutions in other spheres,[114] the central Glav-Textile apparatus tried to solve its own shortage of bookkeepers by luring them away from the Bookkeeping Sub-Department of its own Trade Department.[115]

In this environment, the creation of over forty group administrations in the first half of 1919 and the establishment of their direct administrative ties to Glav-Textile were only preparatory steps. As Lebedev frankly summarized the situation in July:

> Our comrades in the local areas, both in the [group] administrations and in the factory committees and unions, do not understand completely the importance of the task assigned them. . . . A consequence of this absence of consciousness among the representatives of the workers is the fact that upon their election to responsible posts many of our comrades placed their own interests above general ones, and there appeared an intolerable arrogance and rudeness in their attitude toward the workers at the looms and machines.
>
> Thus, for example, one [group] administration, immediately upon its election, firmly and resolutely approached its work with a clear understanding of the tasks required; another went through a period of some indecisiveness and undertook its work only after the *Glavnoe pravlenie* exerted definite pressure on it. One can name a whole series of such administrations.[116]

As one group undertook plans to improve the life of the workers,

> another, often created no more recently than the first, did nothing, as if expecting time to take care of the work assigned to it.
>
> The very same thing is observed in the establishment of relations with local organizations, especially with the factory committees and trade unions. One administration, from the first days of its

existence, carried out its work in close collaboration with them [while] others, unfortunately, stood aside, having themselves forfeited their sources of help and support.[117]

Results were, to say the very least, inauspicious.

There were, to be sure, some model successes, but records bearing out Lebedev's assessment of the situation are far more numerous. On the positive side, a local union in Tambov organized a functioning *kust* on its "own initiative and without corresponding instructions from the center."[118] The March 19–June 12 records of the production group of Petrograd flax–hemp–jute factories present a textbook example of correct procedures,[119] and Lenin himself praised the Danilov group in Moscow for the efficiency of its 1919 operations.[120] More commonly, however, the conditions noted by Lebedev prevailed. Conference resolutions and routine reports from the periphery made the shortfalls of management a recurrent focus and complained regularly of the vulnerability of workers with ties to the countryside to anti-Soviet agitation.[121] One indication of the limitation of operations is the fact that the Soviet scholars who have studied the *kusty* in detail record only a single attempt in 1919, despite the existence of more than forty production groups, to expand a *kust* into part of a state trust as originally planned.[122]

In some cases, the absence of a few key local officials could halt operations. When Lebedev visited a group administration meeting in Alexandrov and questioned the scant accomplishments of the past two months, the official in charge of technical operations answered that work all but stopped when he fell ill with typhus.[123] In a similar case, the Administrative Sub-Department of the Glav-Textile Wool Department could not work effectively for the entire month of April 1919 because its head contracted typhus and his assistant spent two weeks away on family business. This left only the female assistant to the secretary to supervise a few clerks and typists, a staff that union records describe as incapable of filling the requests of factory managements, paying the workers, and dealing with routine problems.[124] A *kust* in Simbirsk delayed its planning work by several months when one member of its technical department died of typhus and two others were injured in a train wreck.[125]

When twenty-six group administrations of textile factories held their first industry-wide conference on March 6–10, 1919, therefore, the proceedings more resembled a seminar on production-group operations than a convocation of knowledgeable officials. Delegates heard reports that duplicated basic instructions already distributed several times through regular administrative channels and widely published. One official, Plavnik, repeated instructions on finance already familiar to those who followed the union press and criticized abuses, such as utilizing operating funds for 1919 to pay wages still owed for the previous year or submitting sloppily compiled estimates. Another speaker, Shishov, provided rudimentary instruction on how to make an estimate correctly. Nogin spoke on basic issues related to establishing a functioning hierarchy: the importance of Raion-Textiles to obey Glav-Textile since they now functioned as group administrations; that local *sovnarkhozy* should not bypass Glav-Textile and issue orders directly to the factories; and the need for the local council of the national economy to subordinate itself to Glav-Textile in areas where it also served as a group administration.[126]

These circumstances gave rise to one additional manifestation of the center's lack of confidence in local cadres: the campaign for one-man management. Soviet institutions had implemented collegial management out of allegiance to the collective propensities of socialism and, of more immediate consequence, as a corrective to the shortage of personnel with a broad range of managerial competence. The idea had its detractors from the outset. The initial references to the need for one-man management appeared as early as the spring of 1918, and additional proposals to eliminate collegial rule followed throughout 1918.[127] The idea of one-man management gained serious momentum in December at the Second Congress of Councils of the National Economy when Lenin proposed entrusting a single person with responsibility.[128]

One-man management received its initial hearing in the textile industry when Nogin raised the issue at the March 1919 congress of group administrations. Undoubtedly anticipating local resistance, the chairman of Glav-Textile carefully qualified his support, professing his loyalty to collegial management and advocating one-man responsibility only in "certain cases." Although Nogin's report went on to cover several other topics, the audience did not miss its chief significance, and one-man management dominated the discussion that followed. The delegate Azarkh raised the most common objection: the director could only be an engineer and, therefore, someone probably unfriendly to the regime and unpalatable to the workers. The delegate Bezmugin countered with what must have been a celebration of the obvious to those present: that conflicts between engineers and workers were not the only ones possible. Factory managements and factory committees also opposed each other in many instances, moving Bezmugin to support one-man management especially in small plants or when individual circumstances merited. The official conference resolution on the matter endorsed collegial management but recognized that in special cases one-man management might be preferable.[129]

The Localities Respond

In contrast to the primacy of organizational questions among representatives of national textile institutions, local officials concentrated on the problems of food, supply, and transportation. Given the demands that local officials heard daily from their constituencies, the food crisis above all dominated factory committee conferences,[130] and individual mills regularly made the fact that their workers had nothing to eat the primary focus of local union communiqués.[131] In the first quarter of 1919, union officials reported factory closings of as long as three to four months as bread rations fell below the subsistence level.[132] The responses of forty-four textile plants employing 73,500 workers to a questionnaire on bread distribution from the union's Statistical Department indicate the severity of the problem: city of Moscow, .55 pound; Moscow province, .14–.53; Iaroslavl province, .60; Riazan province, .17–.91; Vladimir province, .21.[133] The replies, moreover, did not indicate whether these were daily rations or, as was often the case, those distributed on alternate days.

The immensity of the problem led to an excess of suggestions and opinions. When A. Sviderskii reported on the food question to the second national union congress, for example, his analysis—as was common at the time—moved in several directions simultaneously. Bemoaning the shortage of bread, he argued that the food question was the most significant on the agenda. More important to him than quantity, however, was the unsatisfactory apparatus for distribution, and his assessment faulted both aspects of the issue. His tendency to assess multiple blame did not end there. He also chastised the peasants for hoarding grain while simultaneously criticizing the absence of manufactured goods necessary to establish barter between town and countryside. In short, he found fault at every stage of the process of exchange.[134]

The inclination of Sviderskii to strike out in many directions at once typified general reactions within the industry. For example, the fact that *VSNKh* created its own three-member commission in November 1918 to study the food question in Ivanovo–Voznesensk and Vladimir provinces[135] did not prevent both the central union apparatus and the Kineshma textile workers from continuing to operate their own food requisitioning detachments in January–February 1919.[136] In a single exchange at the Second Congress of the Union of Textile Workers, Rudzutak argued that the union should leave food procurement to the Commissariat of Food, Rykunov countered that the All-Russian Council should expand its activities to include the supply of food, and the local delegate Pugachev asserted that the workers, through their group administrations, should initiate their own actions.[137] One solution, begun independently by workers at a number of different mills, was simply to plant food on factory land.[138] On February 15, 1919, a *Sovnarkom* decree legitimized this already widespread practice, and on March 1 Glav-Textile created its own Agronomy Department.[139] The Egor'ev section of the union unabashedly reported that between mid-April and mid-May it devoted most of its energies to trying to purchase potatoes.[140]

Shortages of other materials and of skilled workers closely followed the food crisis in importance since, as the Petrograd Administration of Cotton Factories bluntly stated, it was pointless to establish norms and plans in factories not regularly supplied.[141] More than 40 percent of the textile enterprises in the Central Industrial Region had shut down by the beginning of 1919.[142] Moscow and Petrograd possessed but 30 percent of their demand for fuel and, depending on the industry and area, from 4 to 40 percent of the supplies needed for fall and winter.[143] In spite of the fact that urban enterprises were better supplied than their rural counterparts,[144] even the important Moscow Trekhgornaia Mill shut down early in 1919 when it exhausted its fuel.[145]

Nogin complained that part of this toll was avoidable. Some factories, he argued, closed for lack of fuel despite their adequate raw materials on hand while others possessed fuel but needed materials.[146] Union reports from Ivanovo–Voznesensk bore out his assessment. In May 1919, the former Skvortsov Factory in Pistsovo possessed supplies for only two weeks but fuel for a year and a half. The Gorbunov Factory in Kolobovo reported a two-week supply of materials and fuel for a year. On the other hand, other *kusty* also within the jurisdiction of the Ivanovo–Voznesensk union—Vichuga, Kineshma, Kokhma, and Ivanov—completely lacked fuel.[147] Breakdowns of this nature obviously cannot be blamed directly on the war emergency but rest on the failures of officials to coordinate fuel distribution effectively. Consequently, despite the formation of production groups, reports of additional closings reached the center from virtually all textile-producing areas throughout the first half of 1919.[148]

As supply problems worsened, speculation in textile materials and goods soared. The Wool Department of Glav-Textile reported in May 1919 that carded ribbon, which formerly sold for 50 kopeks an arshin, commanded prices of 200 and even 250 rubles. Aniline dye, usually 30 rubles per pud, now cost 5,000–6,000 rubles, while cane extract, formerly 8–12 rubles per pud, now sold for 5,000 rubles. Department recommendations included ending the toleration of the purchase of black-market supplies by state factories,[149] but such words were easier to pronounce than carry out. The all but total collapse of real wages among the low-paid textile workers moved *Sovnarkom* to authorize the distribution of a ration of flax through the Union of Textile Workers in March to help establish an exchange for bread.[150] By May, the union council took note of numerous complaints that textile workers had not received goods since the tightening of controls over distribution in May 1918. In response, the union leadership authorized a limited distribution of products to the workers.[151] Such measures counteracted worker grievances in the short term but undermined planned distribution.

Social welfare resolutions also failed to produce strong results. A. L. Riazanov complained to the second union congress that a law of October 31, 1918, that extended the right to medical attention to the population had no effect on conditions in the textile industry and that organs for its implementation did not yet exist.[152] Reported deaths—surely lower than the real total—from typhus, Spanish influenza, and smallpox in the Central Industrial Region reported to the Statistical Department of the union rose from 96 in December 1918 to 319 in the following March. At the Moscow Dedov Mill, resolutions concerning education meant nothing because the school was being used as a typhus barracks. A sampling of working juveniles in the Central Industrial Region learned that only 36 worked the four hours prescribed by Soviet law. Another 596 worked a six-hour day, and 66 remained on the job a full eight hours.[153]

The state and union programs to improve factory hygiene and sanitation were also conspicuous failures. The union section in Podol'sk resolved "to take more energetic measures" in March, and when each factory set up its own Sanitary-Hygiene Commission conditions improved, especially where medical personnel were available.[154] Far more numerous than the cautious optimism from Podol'sk, however, were the reports of worsening conditions. The Central Committee of the Union of Textile Workers publicly criticized the fact that the local unions did nothing to clean factories and dormitories, even in the well-organized Ivanovo–Voznesensk branch.[155] At Iakhromo, one union inspector described how rapidly typhus spread through crowded living spaces, where poor ventilation further weakened already emaciated workers and barefoot children. In addition, the use of candles and icon lamps to illuminate congested barracks presented a serious fire hazard.[156] When I. I. Kutuzov toured the Molitov Factory in Nizhnii–Novgorod on March 26, he found 600 workers instead of the former 7,000. Whereas the Molitov had supported the Bolsheviks militantly in 1917–1918 and boasted a party staff of 200, even routine union work had ceased. Dust hung in the unventilated air "like smoke" while workers slept "like ants" in unlit halls where Kutuzov felt sure they would perish in the event of fire. Only in the existence of a nursery, where "a simple hearted woman loves and looks after the children as if they were her own," could he find something to praise.[157] At the former Nosov Brothers Factory in the Blagusha–Lefortovo region of Moscow, wool arrived full of parasites. Workers possessed no protective clothing, and the plant lacked both ventilation and bathing facilities.[158]

Most depressing of all to the center were the charges of bureaucratism, venality, and indifference. From Iur'evets came the charge that whole families endured "dirty, smoky rooms while the comfortable buildings are occupied by Soviet institutions and Soviet employees."[159] On December 8, 1918, the party disbanded the local union leadership in Orekhovo–Zuevo for professional transgressions as well as stealing furniture and occupying too much living space. The union officials offered no defense other than to present countercharges, including that the local soviet used money collected for food to pay its officials.[160] In a separate case, the union instructor Bogachev reported that officials attending a conference in Tver province on February 24, 1919 showed no interest in any aspect of the proceedings except the division of authority.[161]

Both unemployment and worker desertion reached serious levels. The union created a Department of Labor Resources (*Otdel respredeleniia rabochei sily*) to reassign the unemployed to agricultural work, transportation repair, peat-gathering expeditions, factory cleaning and remodeling, mobilization for the Red Army, and other tasks.[162] The highest objective of the union Central Committee was for such workers to repair factories so that production could resume once the supply situation eased.[163] It therefore ensured that those who performed such duties received full pay for two weeks and half pay thereafter and, in addition, ordered the retirement of older workers and invalids.[164] In June 1919, the previously decreed state reinstitution of labor books, intended to restrict workers' mobility and to curtail their frequent changes of employment, was published nationally.[165]

Army mobilizations absorbed a portion of the unemployed but at the price of removing scarce skilled workers from the textile factories. On April 30, the All-Russian Central Council of Trade Unions resolved to mobilize 10 percent of total union membership, in addition to the government levy, for the fighting against Kolchak. The Central Committee of the Union of Textile Workers accepted the task enthusiastically, in addition to organizing its own labor projects and food-requisitioning detachments, and in localities such as Podol'sk mobilization resolutions passed unanimously. In the final accounting, the textile workers of Moscow *Oblast'* sent 10,000 workers in excess of their quota to the front, which enhanced the textile workers' reputation as strong and radical supporters of Bolshevism.[166] Not all textile workers, however, shared either the patriotism or the desperation that drew their coworkers in the Red Army in such numbers. The rapporteur Lomtev noted "a massive exodus . . . of men of military age" from the Podol'sk Troitskaia Factory when the Commission of Assistance to the Mobilization undertook its work, and the Tregubov, Steklovskaia, and two Vaskakov textile factories reported similar flights.[167]

The propensity to flee hunger, unemployment, filth, and army induction fed the already popular perception that textile workers with ties to the countryside were self-sufficient, which the union characteristically countered with multiple measures, especially moral suasion. *Tekstil'shchik* both fed the myth of the self-sufficient worker-peasant and featured appeals to discourage desertion, commonly with analogies to the Bolshevik setbacks of mid-1917.[168] Of major figures, only Nogin publicly challenged the value of the tie to the village:

> Owing to the hostile attitude of the middle peasant toward the proletariat, the worker will be disappointed with the quality of village life. It is doubtful that he will find his livelihood there. What is more, having lived a full century in the factory, having come into close contact with the culture of the great industrial center, he cannot live long outside it. Thus, sooner or later he will return to the factory.[169]

As in the previous year, this heavy reliance on public declarations derived from the fact that cultural–educational work, ideally the instrument for shaping attitudes, continued to founder seriously. The union by no means retreated in principle. It continued to list education as a leading priority and to stress its link to the ability of the working class to regulate production and distribution.[170] Implementation problems, however, were in large measure circular. Assertive local activists suggested that data on demographics and literacy were needed for the entire Soviet Republic in order to attack the problem rationally. The level of literacy among their counterparts in other locales, however, often lagged to the extent that the union could not conduct such statistical work. Local rapporteurs themselves regularly forwarded imprecise, incomplete, and illegible data to the center.[171] Among the textile labor force itself, illiteracy persisted. In March 1919, 18,904 workers remained in 27

Petrograd textile plants, 4,982 men (26 percent) and 13,932 women (74 percent). Of these, 3,312 (66 percent) men were literate, 679 (14 percent) illiterate, and 991 (20 percent) could not be classified. By contrast, only 5,920 women (42 percent) were literate, 4,582 illiterate (33 percent), and 3,430 (25 percent) not determined.[172] Reports from A. A. Ammosova, the member of the union Central Committee who headed its Cultural–Educational Department, indicated that in her view the Petrograd situation typified conditions throughout the industry, and in her experience a low literacy rate usually accompanied a heavy concentration of women.[173]

Programs directed to those already literate fared only slightly better but produced a few conspicuous successes. In December 1918, union cultural–educational departments in the Krasnaia Presnia and Lefortovo sections of the city of Moscow and in several major factories—including the Trekhgornaia, Govard, Giubner, Butikov, and Pervalov–Rybakov—established workers' clubs, libraries, and technical courses. In keeping with the idea that educational work needed to be popularized, workers' clubs also opened in 1918–1919 in Riazan, the Morozov region of Tula province, the Kuvaev (renamed October Revolution) Factory in Ivanovo–Voznesensk, and the Efremovo Factory in Ivanovo–Voznesensk province. Workers' troupe entertainments took place at Efremovo in Tula, and the "Rabochii" library opened in Tver.[174] On January 25, courses on cloth, weaving, and dyeing enrolled 129 in order to combat the shortage of skilled personnel,[175] and the union extended such courses to closed factories later in the spring.[176] Aiming at a higher level of general education, a workers' university in Tula began offering a social science and natural science–mathematics curriculum in February,[177] and 850 studied chemistry, zoology, and physics at the People's University in Orekhovo–Zuevo.[178]

While impressive, these examples do not represent the usual experience. As Ammosova frequently noted, outside the reach of her department in Moscow professional courses fared as poorly as the literacy program. According to the Technical and Production Department of the Administration of the Petrograd Cotton Factories, such projects foundered due to the absence of necessary teaching materials, a shortage of instructors, and a general lack of interest. Although Savel'ev of the Moscow union described the progress of such courses in the capital in glowing terms to Petrograd officials on July 12, expressions of serious doubt greeted his remarks. As usually happened in such encounters, discussants recognized the need for the programs but lacked the means to carry them out. They concluded, as was typical, by agreeing to set up courses in their respective localities "as soon as possible."[179]

At the lowest administrative levels and in the factories, the establishment of national coordination lagged. The lack of a developed Bolshevik presence in major institutions precluded the extension of party discipline to state organs. Even in *VSNKh*, for example, party members dominated the presidium but constituted only 49 percent of the collegia of the various *glavki* and centers. The representation of party members and sympathizers fell to 10 percent among the heads of *glavki* departments, and as late as December 1, 1920, Communists held only 19 percent of the approximately 900 most important *VSNKh* positions.[180] On January 2, 1919, the presidium tried to increase its ability to direct *VSNKh* by creating an internal inspectorate[181] but with scant improvement in systematization. In March, all but thirteen factories ignored the thirty-four questionnaires distributed by the Glav-Textile Department of Coarse Cloth, as they did all other communications from the department.[182] In May, the Glukhov Factory in Tambov province habitually ignored Glav-Textile directives without apparent repercussions.[183]

Poor communications bore part of the responsibility for this situation. Telegrams required a week to reach their destination,[184] and in some cases ignorance of the desires of superordinate bodies rather than petulance lay at the root of disobedience and noncooperation. The editors of *Tesktil'shchik* were discouraged to learn, when they polled local areas at the end of 1918, that in many areas no one had heard of the journal.[185] Remedial steps frequently misfired. In May 1919, for example, the Glav-Textile Department of Coarse Cloth delegated additional responsibilities to the group administrations in order to reduce its load of paperwork,[186] and the union in Ivanovo–Voznesensk employed a similar strategy of devolving work upon its departments and subdepartments. Korolev reported with chagrin, however, that the program in Ivanovo–Voznesensk led only to greater local separatism.[187] Ubiquitous rumors also eluded control. In one instance, several newspapers in

Tambov led workers to expect a 50 percent pay raise. When the union achieved only 20–30 percent, and for some jobs 15 percent, the local union lost credibility.[188]

Conclusion

In the final analysis, the form of centralization that emerged in October 1918–July 1919 was limited largely to a concentration of decision-making authority. If prerevolutionary Bolshevik expectations had combined central direction with local initiative, the non-materialization of the latter resulted in a greater reliance on the former, given the range of choices presented by the party leadership's understanding of Marxism and the history of the party itself. Yet the center lacked the ability to transform the concentration of decision making into the centralization of actual administration in the textile industry, and lower officials were unwilling to surrender their prerogatives voluntarily. Under existing circumstances, neither central nor local institutions possessed the resources to alter material conditions and thereby increase their influence and credibility.

For the long-range achievement of revolutionary objectives, the ungovernability and mismanagement of the portion of the industry that remained in operation presented problems more serious than even the immediate material pressures. Technical personnel and the conscious and mass workers who filled local offices often lacked the necessary skills or displayed little desire to follow central directives unquestioningly, as shown by the union's second national congress. They frequently formulated their own policies and displayed proprietary attitudes toward their jurisdictions. This created a rift not only between local and superordinate institutions but between lower organs and their immediate constituencies as well. The fact that supplies did not circulate efficiently even among neighboring factories linked to the same production group, for example, cannot be reasonably read as antagonism either to the Bolshevik Revolution or to the revolution in the abstract but as a demonstration of the full sway of localism.

To these rivalries among the constituency of the revolution one must add the manifestations of class conflict. This chapter has demonstrated that in this regard aspirations remained remarkably consistent. As the form and content of local communications have repeatedly shown, the main pressures exerted by mass workers on local officials continued to gravitate toward goals as old as the prerevolutionary textile strike movement: material betterment and the redress of grievances against class enemies. Therefore, and not to put too fine a point on the issue, the recurrent postrevolutionary calls for material relief and the further expropriation of the oppressors were grounded in the principal goals of the revolution, as understood by the rank and file in the textile industry. Beyond this, local officials joined the rank and file in their unwillingness to trust former owners and technical personnel, as evidenced by the opposition of local officials to one-man management despite the demonstrated shortcomings of collegial administration. For their part, upper union officials showed signs of tolerance of one-man management, but as a practical consideration rather than an endorsement of the former owners and managers. Indeed, the determination with which the union leadership wrested adminstrative authority away from factory owners by eliminating Centro-Textile testifies to the full extent of class-based animosity.

In sum, during October 1918–July 1919, responsible institutions and their constituencies found themselves confronting the same range of problems they had faced immediately following the October Revolution. Now, however, the correctives formulated took into account a full year of revolutionary experience. The result brought the revolution out of its stage of unbridled experimentation, but in a manner that deepened rather than alleviated existing points of tension.

5

The New Crisis: Victory and Rising Expectations

From the middle of 1919 through the end of 1920, the Soviet Republic experienced conflicting fortunes. A stronger and better-organized Red Army established its superiority over the disjointed White forces in the second half of 1919 and did not yet foresee the hostilities that would break out subsequently against Poland. By April 1920, only the troops of Baron Wrangel remained at large, and even these lacked general support in the areas where they operated. To a population that had experienced continuous warfare since 1914, rising military fortunes could have no effect other than to engender hopes for better times, and by early 1920 pronouncements of confidence began to appear regularly for the first time since 1917.

The prospect of victory, however, also generated new and different pressures on the state and party leadership. Society as well as the general party membership began to voice increasingly impatient hopes that the material crises would end and that the regime would provide tangible evidence that the awaited transition to a more equitable socioeconomic order was under way. But rising expectations neither fed the population nor revitalized industry. The impending end of the war actually undercut the legitimacy of some of the measures on which the state had come to rely, particularly grain requisitioning and the extension of nationalization even to small-scale enterprises. Moreover, demobilizing a victorious army was certain only to complicate some contemporary problems, especially by increasing the ranks of the unemployed.

In a large sense, therefore, victory brought with it stress more intense than that of the military struggle. Without trivializing the suffering of the Soviet people during the civil war, one can say that the war at least provided a rallying point for mass mobilization. The Whites presented a palpable threat to the security of the population, and the defense effort represented the common cause of protecting the revolution from the reactionary intentions of the Whites and their foreign supporters, as the Bolsheviks presented matters. Once this threat began to recede in the public consciousness, the state was called upon to assume greater responsibility for the welfare of society, and without the use of its most recurrent previous tactic for addressing the population, the emergency campaign.

The party and state reacted with measures that appeared to grasp at establishing any kind of direction whatsoever in what was becoming a completely unmanageable situation. In rapid succession, the government implemented new programs or accelerated those already in motion—an even greater centralization of cadres, the transition from collegial to one-man management, the use of piece rates to raise productivity, the creation of so-called shock factories—as an alternative to the faltering production groups and as a way to utilize resources more efficiently. Local officials and workers, however, resisted centralized direction as stridently as before. Nominally subordinate institutions continued to perform largely as interest groups as soviets, *sovnarkhozy*, unions, commissariats, and other organs competed with one another at each administrative level. Within each institution, lower organs resisted subordination to their immediate superiors and to their central parent organization. Most, as we have seen, had not been organized effectively and incorporated into their respective apparatus in the first place.

The large number of the local officials who championed local independence did not necessarily act on behalf of their constituency but frequently out of self-interest. By 1919–1920, there could be no denying the gulf that separated local officialdom from the rank-and-file textile workers. Thus, at the moment when military victory presented new socioeconomic possibilities, local officials exploited their positions for personal enrichment and textile workers displayed a general indifference even to the overtures of their own trade union. Where workers even knew that a national union existed, they did not cooperate with its efforts to organize them. If enrolled automatically at their factory, they abstained from active participation in union affairs. The disappointed expectations of a better material life had in three years led not to the crystallization of a new society but toward continued fragmentation.

New Prospects, Old Problems

The military success that gave rise to an elevation of expectations began with pivotal campaigns in 1919. By March 1919, Kolchak, the so-called dictator of Siberia, crossed the Urals and began his drive to central Russia; Iudenich gathered an army in Estonia and reached the outskirts of Petrograd; and Miller, with very limited supplies, made his greatest gains in the northeast, the least important and most ephemeral theater of the war. Finally, with the Reds concentrating on holding back Kolchak in Siberia, Denikin mounted a new offensive from the south with the objective of taking Moscow (he announced his intentions in the Moscow Directive of July 1919) and at the same time conquered much of the Ukraine. By summer 1919, however, the revival of the military under Trotsky's direction began to produce dividends. Both campaigns in the north were won. The Red Army also drove Kolchak back across the Urals, and by October he was in full retreat through Siberia. This allowed the Reds to transfer troops to the southern front, and Denikin was halted 400 miles from Moscow. By the beginning of 1920, all White armies were in retreat (Kolchak was eventually captured and executed; Denikin and much of his force were evacuated abroad). Wrangel subsequently pursued a rearguard action, and in 1920–1921 the Red Army would attempt an ultimately unsuccessful Polish campaign, but by the spring of 1920 public perceptions were that eventual victory in Russia was assured. The war was not so much won by the Reds as lost by the Whites. They organized even less effectively than the Bolsheviks; received too little aid from the Allied intervention to influence the outcome of the war; failed to establish significant mutual cooperation among White leaders; refused to address social reforms and indeed reinstituted former conditions in the territory they conquered; lived off the population while advancing, which caused problems in retreat;

and placed Great Russian nationalism over pragmatism, which alienated minority nationalities.

Unfortunately for the Bolsheviks, general conditions did not improve in 1920 despite successes in warfare. Contrary to Lenin's earlier expectations, social and economic change followed their own dynamics and did not respond directly to alterations in institutional forms. They certainly did not react immediately to upturns of political and military circumstances. While this appears self-evident, especially from hindsight, it was in fact one of the major lessons of the first years of Soviet rule. Creating a new order turned out to be considerably more complex than taking over banks and organizing industrial trusts supervised by representatives of the working class. By 1919–1920, the condition of the country had deteriorated too far for military victories alone to revive it. Thus, the fact that production levels fell below even those of 1917—in the national economy as a whole and in the textile industry in particular—largely overshadowed military achievements. Previously weak labor discipline was reported to have disappeared almost completely, the organization of new and reliable administrative bodies lagged seriously behind the plans projected in the first months of Soviet rule, and the workers' living conditions deteriorated to new depths. Ironically, therefore, the party's economic guidelines of 1917 did not face their greatest challenge in 1918–1919, when the very survival of the regime was in dobut, but when the Soviets later won the opportunity to create their own society.

With pressure to produce tangible gains mounting, the leadership actively pushed its own version of what was taking place. When the Eighth All-Russian Congress of Soviets met in December 1920, Rykov outlined economic goals that were identical to many of those put forward three years earlier: improving the food supply; revitalizing industry and transportation; using materials rationally; organizing the labor force; formulating a single economic plan for the nation; resolving institutional parallelism and bureaucratism; developing further local independence of action; and drawing the working class into economic life.[1] Rykov, foreshadowing what was to become a recurrent theme in Soviet and Western historiography, placed most of the blame for the lack of economic progress on the conditions of the civil war. With specific reference to the textile industry, he explained: "In the textile regions all industry almost died [*zamerla*] from lack of cotton in the past year." With new deliveries on the way to the Central Industrial Region, Rykov was optimistic of the future.[2]

Actions surrounding the congress, however, showed that Rykov's explanation evaded the important question of local support. Serious challenges to centralized authority materialized within the upper reaches of the party in late 1920 to early 1921. These were mirrored in society by an upsurge in the intensity of strikes and disturbances, culminating in the major symbolic setback of the Kronstadt revolt of March 1921. There was no shortage of proposed correctives. The Democratic Centralists in the party advocated a revitalization of its apparatus, especially at lower levels, through a delegation of greater responsibility to noncentral party organizations. The Workers' Opposition championed greater freedom for the unions (at least at the level of their respective national leaderships) from party direction and fundamentally questioned the dictum that independently functioning unions were a contradiction in a workers' state. Although these positions existed as early as 1918, both became more intense in response to increased efforts to concentrate authority at the top of the party and state apparatus in 1919–1920, as in the further subordination of the soviets to the party and Trotsky's proposal to extend military principles to the organization of labor. The leadership actually continued to increase central direction in the administration of the economy through the very eve of the Tenth Party Congress of March 1921.

The task facing the Soviet Republic at the end of 1920 was to revive the economy and rectify overcentralization without losing authority and, equally important, violating its ideological orientation. At the Tenth Party Congress, the principle of party direction prevailed, bringing the defeat of the major opposition groups. Lenin and his supporters were not completely intransigent, however. Their short-term solution was the New Economic Policy (NEP) adopted at the congress, which included the partial reintroduction of the market and the denationalization of some small-scale industrial enterprises. While not, as some have argued, a direct return to state capitalism, NEP—carrying a commitment that it would remain in effect in the foreseeable future—did reintroduce the idea of a gradual transition to socialism.

With specific reference to the textile industry, however, local discontent transcended the party congress. Kutuzov, chairman of the textile union, reported (after the fact) that the union leadership linked the decentralizing proclivities in evidence at the Eighth Congress of Soviets to the apparent complete failure of local administration. In February 1921, the union tacitly admitted the inadequacy of the existing economic apparatus when it dissolved the group administrations.[3] By September 1921, when the Fourth All-Russian Congress of the Union of Textile Workers met, Kutuzov spoke with even greater candor. He stated that in 1918–1920 "we felt a complete obstruction [*zakuporka*] to our work in the local areas" and that "the *glavnoe pravlenie* had already begun to turn into a so-called bureaucratic apparatus owing to the fact that it never undertook any positive acts for the good of the factories."[4] Such admissions attracted scant sympathy from the rank and file. Instead, on September 11 local officials and supporters of the *sovnarkhozy* devoted an entire session of the union congress to venting their animosity toward the union leadership and Glav-Textile about the ineffectiveness of union policy, irresolute leadership, and a general disregard for the local areas.

New Economic Construction

In the textile industry, optimism generated by military successes manifested itself in a new stock phrase in the winter of 1919–1920—the beginning of economic construction—that became standard in the records and resolutions of congresses and conferences. Local reports adopted its tone,[5] and on March 19, 1920, Gorshkov of the Central Committee of the union greeted a conference of factory committees with the prediction that the Reds would win the war and an announcement that the textile industry would now be reconstructed.[6] Such rhetoric reached a plateau at the Third All-Russian Congress of the Union of Textile Workers, held April 16–20, and from there union journals and communications disseminated it widely in the industry. When Lenin addressed the congress, he stressed that peaceful reconstruction of the economy must begin immediately and that the time was right for the peasants to rally behind the workers's government: "At present we are unable to give them manufactured goods. Here we must succeed in approaching the peasantry and making clear that in the conditions that we are now living through they must give their bread to the workers and be reimbursed later [*v ssudu*]."[7] Lebedev, the featured union speaker, emphasized similar themes, especially that the recovery of the national economy and of the textile industry had begun.[8] The first of the resolutions passed at the end of the congress consequently listed the organization of the national economy as the top priority of the country following the defeat of the Whites. *Tekstil'shchik* published the resolutions in its next issue and began to feature related articles that reiterated the idea of the beginning of a new era.[9]

Implementation proved problematic. While virtually all those sympathetic to the Soviet state declared their support for establishing coordinated state regulation in industry, institutional loyalties and conflicting views on how far measures of economic expediency might deviate from preconceptions of revolution became even more important between July 1919 and December 1920. Organizing industrial administration was a task as old as the Soviet regime, and it commanded the main attention at so many conferences during this period[10] that it might prove more difficult to discover gatherings that did not feature the issue prominently. Experience, however, had also taught the limits of discussion alone. Certainly the Second Congress of the Union of Textile Workers had given the issue of organization renewed emphasis in early 1919, but the need to devote continued and overriding concern to the topic indicated that the broader plans for the textile industry had not been communicated to all relevant local areas even by the end of the year. As Nogin had argued before a conference of *kusty* in September 1919, forming additional group administrations would accomplish little unless the *character* of their work changed. Circumstances demanded not simply the formation of organs but the implementation of the actual process of administration.[11] When the Seventh All-Russian Congress of Soviets met December 5-9, 1919, therefore, the question of organization came again to the forefront and featured the full range of opinion accessible to the politically attentive public.

General objectives and institutional loyalties clashed repeatedly at the congress. On December 8, 1919, T. V. Sapronov, in a speech to the Seventh Congress of Soviets that foreshadowed a stand he would take on Democratic Centralism at the Tenth Party Congress in 1921, candidly addressed various shortfalls of the attempt to centralize operations:

> Comrades, the organizational principles of the creation of soviets both as political organs—organs of power—and as economic organs were expressed by the Soviet Constitution, but expressed with insufficient clarity. In the process of creating soviets, the experience has followed various paths. On the one hand, local soviets often considered themselves the completely autonomous government in the local areas, considered that they were subordinate to no one and existed independently; on the other hand, the tendency of the central organizations followed the opposite path, that is, the elimination of any power in the local areas. The organization of the national economy, the organization of a whole series of departments began to bring about central commissariats and *glavki* not as departments of executive committees [of soviets] but as departments of People's Commissariats, *Glavki*, and Centers.[12]

Ideally, he argued, provincial executive committees and city soviets should carry prime responsibility for the implementation of central decisions and other organs should play subordinate roles.

The problem, in Sapronov's view, was that the commissariats, *glavki*, and centers hindered organization by making their own unmanageable organs parallel to the soviets:

> These departments, located in the province or district, are not subordinate to the [soviet] executive committee and are cut off from the center. Thus, instead of centralized contact with the provincial and district economy our economy on an All-Russian scale is divided into a whole series of axes [*stolbikov*; literally "columns"] that are in no way connected among themselves and which perform parallel tasks.[13]

VSNKh, according to Sapronov, was the worst offender because it tended to become stronger, as he viewed matters, as the number of its various organs proliferated. From the moment of their inception, he held, councils of the national economy directly opposed the authority of the soviets at the provincial, district, and town level, but without providing a working alternative to the ideal of soviet rule:

> Moreover, within the provincial councils of the national economy themselves there is no single organization but tens of independent

units: Glav-Leather, Centro-Textile [*sic*], Glav-Paper, and other *glavki* create provincial and district departments that are not only not subordinate to the Central Executive Committee but also are not subordinate to the Presidium of the Provincial Council of the National Economy. In the provinces, there is no centralized apparatus which could unite the whole provincial economy; on the contrary, there is a whole series of independent [organs], not connected by mutual organizational ties, often contradicting each other . . . [and] performing parallel work.[14]

His solution was to unite management authority under the soviets at all levels,[15] but he lamented that the institutions involved could not agree even on the delineation of administrative boundaries. Sapronov ruefully noted that "the Council of the National Economy chooses one boundary and Centro-Textile [*sic*] is marked by other boundaries distinct from provincial and district soviets, etc."[16]

Given Sapronov's propensity to criticize the soviets' internal structure as well as all major institutions competing with the soviets for influence, the discussion that followed his pivotal speech reflected a broad range of institutional and programmatic loyalties. In a prelude to debates that would further divide the Bolshevik leadership in the future, the influential V. Ia. Chubar' critized those who, like Sapronov, opposed centralizing the national economy. M. I. Kalinin, who had become chairman of the Soviet Central Executive Committee in March 1919, seconded the sentiment by stating that "the development of the idea of communism itself leads to the concentration of the basic means of production. Whoever opposes this opposes communism and becomes a petit-bourgeois, a Menshevik and so on."[17] Osinskii, the Left Communist, then criticized everyone who had preceded him to the floor. He found fault with the relationship of central and local soviets but noted that all soviets united at least in their unyielding opposition to the *glavki* and centers, to the detriment of national coordination. He chastised Kalinin specifically as a central official who spoke on matters in which he had no personal experience. Finally, he conceded to Chubar' that specialized *glavki* were necessary, but he denied them the right to become independent entities within the economy. Providing yet another perspective, L. M. Kaganovich denounced the entire discussion for treating the *glavki*, in his view "the most burning question of the workers in the local areas, in only a demagogic way."[18]

Nogin took the floor later that night to present the case for Glav-Textile. Previous speakers, he asserted, had erred in not taking economic preconditions into full account in their discussion of organization. Only *glavki* such as Glav-Textile, he maintained, possessed the scope of operations and relevant experience to procure and disperse the short supplies of cotton and arrange the transportation needs of the textile industry on a national scale. Provincial executive committees of soviets or councils of the national economy operated too narrowly to resolve such problems, in his view. The present state of the economy, he concluded, dictated that officials of the local soviets not interfere in industries where they had no expertise. He felt that centralization, not decentralization, must be the guiding principle of policymaking, but it was clear that Nogin meant for the individual *glavk* to be the main centralizing agent.[19]

While expressing opposing views on the basic philosophy of administration, such discussions managed to talk past a more fundamental issue: the ongoing shortage of talented and cooperative cadres, without whom neither centralized nor decentralized management could take shape. Group administrations continued to complain both of personnel shortages and of the scant opportunities to attract new cadres,[20] while central officials decried the poor performance of those presently on the job. Lebedev publicly decried apathy, poor discipline, and low productivity "not only among the part of the workers that lacks a high consciousness, but frequently also among the responsible leaders of both the trade unions and organs of administration."[21] Moreover, many qualified personnel on hand preferred work in other industries even to employment in Glav-Textile. As the 1919 annual report of its Administrative Department explained, the fact that Glav-Textile paid only money wages rather than goods particularly caused resentment among employees (*sotrudnikov*), who found service so unpleasant that they sought any excuse to be relieved of their obligations. Glav-Textile took stern measures to prevent personnel from leaving their posts, but as a result in disciplinary cases the prospect of "dismissal does not appear to the workers as a punishment, but, on the contrary, the desired result."[22]

Plagued by shortages, central and local organs competed with one another for the services of trained cadres. Factory closings drove union membership down, thus contracting the pool of potential talent and weakening the union generally. As previously noted, membership stood at 484,747 on July 1, 1919, already reduced from 642,518 on January 1, and it fell to 432,740 on November 1.[23] Meanwhile, the problem of mustering a full complement of the union Central Committee continued, even after the third union congress elected an expanded new body of fifteen in April 1920. As early as one month later, *Tesktil'shchik* reported that only seven members worked regularly, and only nine attended the pivotal meeting of May 15 at which spheres of responsibility were assigned.[24]

The union leadership had taken the position in 1918 that central regulatory bodies should be allowed to co-opt qualified personnel from local organs, and it in no way retreated in 1919–1920. Local organs understandably reacted by energetically trying to retain whatever talent was available to them. A few local bodies, including the consistently cooperative Danilov Mill, acceded to co-optations of personnel without protest,[25] but such instances were apparently not the majority. On February 5, 1920, in fact, I. I. Kutuzov felt the need to reiterate to the Moscow Provincial Conference of the union "that it is necessary to draw to the leadership of the union the most experienced workers from the local areas."[26] Two months later, a resolution of the Third Congress of the Union of Textile Workers declared: "The congress expresses its approval of the [Union] Central Committee and suggests that the new C.C. decisively and persistently utilizes its right to recall workers from the local areas for work in the center."[27] This certainly did not decide the issue, and local organs continued to resist. On September 20, the Moscow Provincial Conference of Factory Committees again had to reassert the right of the provincial union to draft the services of the best local workers.[28] In the following month, Kutuzov advanced the case of the center yet again in *Tesktil'shchik* when he cast doubt on the common response to calls for cadres: none were available. In spite of the many army mobilizations, he argued disingenuously, personnel would exist for union work and service in economic regulation if only the workers would select them from among their own ranks.[29]

By 1919–1920, however, even this center–local focus, guided as it was by prerevolutionary aspirations, was largely out of date since accumulated experience since 1917–1918 had created an impetus of its own. In the late stages of the civil war, the functioning organs— however imperfect—had defined their own agendas and, more to the immediate point, had learned to protect their institutional inter- ests both laterally and against the center. In short, while Bolsheviks at the center debated competing philosophies of organization and sought cadres, officials at other levels functioned according to a perspective of their own. One may be sure that they in no way viewed their own tenure in office as a temporary expedient.

If we examine the stenographic record of the Congress of Group Administrations and Gub-Textiles held September 16–22, 1920, we can see different dimensions of the situation. As Lebedev observed at one session:

> The problem of the relationship among the various organs which carry out the business of the textile industry for the Central Man- agement is on the agenda because almost daily a whole series of communications arrives from the local areas about all kinds of misunderstandings, of clashes and conflicts between local organi- zations—between factory committees and the managements of various factories, between factory committees and group adminis- trations, and between factory managements and group administra- tions. There is also a relative lack of clarity about the correct relationship that must exist between the Central Management and the group administrations. And often the group administrations in the local areas carry out decisions independently which, it turns out, they should not have carried out.[30]

Had his speech ended there, it would not have differed from count- less others, but Lebedev probed further:

> It is necessary to say that in spite of the fact that it is in the group administrations that the standard for the others is set and an example in all dealings provided, there are not only technical personnel but workers who hold not a proletarian view but the viewpoint of the small-time plunderer who tries to get everything for himself and spits on everyone else.[31]

The discussion that followed indicated that Lebedev had trod on numerous allegiances.

The debate generated both heat and conflicting perspectives. Before Lebedev even finished speaking, a voice from the floor chastised him for not suggesting solutions. Not waiting for that issue to take shape, the chairman of the Egor'ev-Ramen Bureau launched a defense of the group administrations. He did not respond to the charges against the attitude of managerial personnel but instead counterattacked Glav-Textile. If directives were not carried out, he asserted, it was only because Glav-Textile failed to outline spheres of authority clearly. In his own production group, four different branches of the union and four *sovnarkhozy* interacted with a single group administration. Moreover, even though the group was located about 100 versts from Moscow, Glav-Textile never sent representatives to check on proceedings or give instructions. How, he asked rhetorically, could a unified production plan emerge? Underscoring the discrepancy between conference discussions and action, he concluded with the charge that "here our talk is democratic, but our actions are bureaucratic." Employing a different approach, the delegate Kolovich from Petrograd denied that any single cause could explain the friction among organs and expressed bitterness that under the existing circumstances Glav-Textile would criticize the way local organs implemented instructions. On the opposite side, Shal'nov defended the Central Management and blamed local organs for a lack of initiative, and in his closing remarks Lebedev gained the floor again to decry the "abnormality" of relations in all institutions.[32]

Two messages emerged from this. First, representatives of the union center, in this case Lebedev of the Central Committee of the Union of Textile Workers and of Glav-Textile, no longer blamed only bourgeois sabotage and the persistence of petit-bourgeois attitudes among workers closely tied to the countryside for existing problems. This exchange candidly admitted that the center considered local officials, many of whom were legitimate members of the working class, as forces opposed to complete subordination to higher organs. Second, in this instance debate over general issues assumed a specific and more pronounced character of institutional partisanship. Whereas lamenting the magnitude of the problems facing the revolution had dominated exchanges of this type from the beginning of Soviet rule, this debate unabashedly failed to address solutions and centered on defending one's own institution from outside criticism.

In practice, therefore, cooperation continued to deteriorate at all administrative levels. The observation made at the September conference of group administrations that group and factory managements placed their own concerns above class interests and operated each factory in a proprietary manner reflected contemporary accepted wisdom.[33] Speaking for Glav-Textile, Bubnov told the third union congress on April 18, 1920 that the center still found itself fighting for the acceptance of its authority over localizing tendencies. Only when local areas accepted central government administration, he noted, could the center support the transfer of real initiative to the local areas.[34]

Words such as these did not alter behavior. On the contrary, the records of factory committees and professional organizations at the lower administrative levels in the Central Industrial Region continued to reflect an overriding concern with immediate, practical concerns. By 1919–1920, gatherings of these bodies touched upon general organizational or ideological issues even more rarely than in the early months of Soviet rule.[35] Reports of conflict between factory committees and managements continued regularly to reach the union leadership,[36] as did accounts of clashes between group administrations and councils of the national economy.[37] Fulminations against bureaucratic waste continued, as when a Moscow provincial conference of factory committees, district unions, factory managements, group administrations, and commissions of female workers (*komissii rabotnits*) indicted the *glavki* and centers as well as "other Soviet economic organs" in October 1920.[38]

Thus, enjoying neither loyalty, enforcement powers, nor systematic operations, the union, Glav-Textile, and government organs continued to resort to redundant directives in a self-consciously futile attempt to influence regional and local behavior. Thus, when the union's Organizational–Instructional Department formulated revised regulations for factory committees in August 1919, it issued instructions that would not have been out of place a year and one-half earlier. In addition to reiterating that the factory committee was the local unit of the union, a point established at the First Congress of the Union of Textile Workers in January 1918, it

offered counsel on the committees' term of office, the frequency of
their meetings, the necessity of filing reports, the mechanics of dues
collection, and related procedures already well publicized.[39] On
October 9, 1919, the Central Committee of the union felt it necessary again to remind *raion* unions and factory committees to hold
monthly meetings and, moreover, of the seemingly obvious point
that orders could be issued in the name of the union only if signed
by persons authorized to do so.[40] In a separate example on December 24, a meeting of the Textile Section of the Council of the
National Economy of the Northern Industrial Region demonstrated
how poorly information was disseminated. Even at this relatively
high administrative level, discussion broke down when two different versions of the resolutions of the recent Seventh Congress of
Soviets emerged, and the accurate version could not be identified.[41]
Examples can be multiplied. On January 29, 1920, a provincial
union congress still found it productive to listen to a rudimentary
definition of a Gub-Textile,[42] and the third union congress in April
included in its resolutions a full explication of the institutional
hierarchy in the industry: what Glav-Textile was, how the union
influenced its composition, a definition of a group administration,
how a Gub-Textile operated, and the procedure for forming factory
managements.[43] Indeed, the Central Committee of the Union of
Textile Workers reported that it had held no fewer than twenty-
three conferences of factory committees between February and
October 1919 that focused chiefly on questions of organization but
with little evidence of success.[44]

One-Man Management and Discipline

The existence of so many sharp divisions on centralization and
organization directly affected the revitalization of industrial production. We have already documented the fact that directives on
raising labor productivity were highly redundant. Policymakers, of
course, did not limit themselves to pious pronouncements, and they
formulated new responses to deteriorating conditions and the collapse of the ruble. Indeed, in 1919–1920 central agencies supplemented public suasion with a series of innovations designed to tie

the workers' remuneration directly to output and to increase the accountability of management.

One controversial proposal that moved to the forefront nationally and in the textile industry at this time was the replacement of collegial administration with one-man management. As we have seen, Lenin had seriously begun to press this issue as early as December 1918, but Nogin had publicly given collegial responsibility a qualified endorsement for the textile industry as recently as March 1919. By autumn, a shift occurred. At the September 1919 conference of group administrations and Gub-Textiles, Kutuzov, the chairman of the textile workers' union, labeled one-man management a "dangerous step" that would remove existing checks on technical personnel involved in factory management.[45] Nogin, on the other hand, began to back away from his earlier position. He still favored the principle of collegiality, he stressed, but a recent conversation with Lenin at a meeting of *Sovnarkom* had impressed upon him the cumbersome character of collective responsibility. Both he and Lenin agreed, Nogin said, that worker participation in the collegial system had been a step forward, but he now felt that small enterprises especially would be better served by the management of a single person. Nogin reiterated that he did not favor replacing collegial rule completely, that to do so would be a mistake, and that the collegial system was superior to one-man management, but his stand was more equivocal than it had been previously.[46] For the time being, however, neither Glav-Textile nor the Union of Textile Workers energetically supported one-man management, and indeed both clearly feared its potential to undermine their authority in the industry.

In this posture, they were uncharacteristically in harmony with the advocates of greater local authority, who already distrusted the individual technical specialists and feared giving them more power. Ongoing reports of activity characterized as bourgeois sabotage left no doubt about local opposition to the rule of technical experts, and one classic example from October–November 1919 illustrates the enduring level of hostility. On October 25, the administration of the Monin (formerly Shishov and Grubnov) Factory decided to discharge a member of the management collegium, one Engel'bert Fridrikhovich Miller. The cloth master Miller, the meeting charged, had tried to expand his authority over his less-experienced colleagues from the day the group took office, and he engaged in theft and speculation. Dismissal alone, however, did not satisfy his opponents, and on November 1 the factory management gave Miller five days notice to vacate his company apartment. When representatives arrived to inspect the quarters on November 5, one day early, Miller's wife initially pretended not to recognize them and refused entry on the grounds that they might be bandits who would carry off her possessions. Her words contained unintended irony. To resolve the stalemate, the Communist faction from the former Belov Factory intervened, gained entry, and examined the contents of the quarters, but with little benefit to the Millers. In a scene that would be replicated many times during dekulakization a decade later, the inspectors fulfilled the worst fears of Miller's wife by seizing twenty-seven categories of goods, including two armchairs, a bread knife, a large mirror, Miller's library and bookshelf, and his gramophone and records.[47]

In short, any suggestion to enhance the role of technical personnel in the plants aggravated already sensitive animosities and fears. To advocate greater authority for specialists appeared to give up the battle against management sabotage, reignited existing factory rivalries, and frightened local regulatory institutions into thinking that they would lose their authority. On a different level, the collegial form of dispersing responsibility just seemed to many the only correct "socialist" approach. At best, one-man management appeared a poor compromise with capitalism, and at worst a complete surrender.

Lenin and Trotsky, nevertheless, continued to push for one-man management within the party and in *Sovnarkom*, and their desires ultimately prevailed. They carried their case to the Eighth Party Conference (December 2–4, 1919) and then to the Seventh Congress of Soviets, where Trotsky argued the proposal and Lenin supported him by citing the shortage of cadres as a rationale.[48] When *Pravda* then printed a number of Trotsky's theses on organizing production and raising labor productivity, however, the Central Committee of the Union of Textile Workers made a point of placing their objections on the public record. On December 18, the union's Central Committee resolved that collegial management was the only path to production on a "communist" basis. Trotsky's position negated collegiality, which in turn disrupted a united workers' front, the union leaders stated. Discipline and productivity could emerge only through the efforts of corresponding workers' organizations, and it should not cause surprise that the union leaders declared the trade union to be the proper coordinating organ.[49] Despite this kind of opposition, the Ninth Party Congress (March 29–April 5, 1920) decided in favor of one-man management after heated debate.[50]

Faced with this accomplished fact, the Bolshevik-dominated Central Committee of the Union of Textile Workers and Glav-Textile reversed their earlier position. While we have no way of knowing what maneuvering unrecorded in written records took place, the public turnabout was dramatic. On April 18, Bubnov spoke to the national textile union congress on behalf of the Bolshevik Central Committee about the determination of Glav-Textile to achieve the central direction of textile production, a proposition Lebedev had placed before the delegates two days earlier.[51] Other delegates presented specific proposals that would have seemed out of place at the previous congress, such as limiting some of the prerogatives of factory committees.[52] In short, Glav-Textile and the union leadership appeared to adhere to party discipline, accepted the decision of the Ninth Party Congress, began directly to implement one-man management, and within three to four months executed the change.[53] In so doing—and this point deserves emphasis—they continued to work for concentrating authority in the industry in their own hands and adapted one-man management to serve this end. On April 14, 1920, Glav-Textile declared its support of "the careful and gradual implementation of the beginning of one-man management" through outlining more strictly the duties of the factory committees and managers and, in some cases, merging the managements of large-scale enterprises with the management of the group administration.[54]

The resolutions of the union congress specified that the conversion to one-man management should begin with single enterprises or small groups and extend from there to the large-group, *raion*, and central levels. In the factories and small groups, a Director–Administrator or Director–Specialist would take charge. If the administrator were a worker with the requisite abilities and experience, his assistant must be a technical specialist. If a so-called bourgeois specialist were director, a commissar was to serve as chief assistant with one or two workers assuming assistant roles as well. Where collegial management remained operative during the transition, its character would change. Each collegium formerly answered collectively, but members would now receive precise, specific responsibilities for which they would be held personally accountable.[55]

The mechanics of implementation in the following months indicated strongly that the union would utilize one-man management to extend its authority, despite its initial opposition to the measure itself. It set the term of service for each manager and reserved itself the right of removal. Glav-Textile was to issue the formal appointments.[56] Moreover, the wording of new directives indicated that the union and Glav-Textile expected future interaction to be more efficient than when dealing with a number of different people under the collegial system. A directive of the union's Economic Department left nothing open to interpretation when it defined full authority and responsibility. The manager would answer directly to Glav-Textile, and all technical and administrative personnel, including his assistants and chief engineer (if not an assistant), were fully subordinate to him. The manager could make any decision within the plant except increasing or decreasing production and closing or partially closing the plant. Only in the absence of the manager could assistants exercise authority.[57]

Between the union congress in April and the fall of 1920, one-man management became the predominant form of accountability in the textile industry, usually with a specialist as director and a worker as his assistant.[58] The union moved gradually at first, and in mid-June *VSNKh* still approved the formation of factory managements according to the old formula of one-third workers, one-third technical specialists, and one-third appointees of workers' organizations.[59] In late June, however, Glav-Textile and the Organizational–Instructional Department of the union instituted one-man management in two of is best-run group administrations: the eight factories of the Danilov Group and the ten enterprises of the Alexandrov Group.[60] When the inner presidium of Glav-Textile met on July 7, it used the formula of appointing a specialist assisted by a worker to alter the management of the Repikhov, Luzov, Ivanov, Bratsev, Kupavin, Naro-Fominsk, Moskvarets, Frianov, and additional factories, and on July 15 repeated the pattern in its appointments for the factories of Ivanovo and Sereda.[61] Appointing managers and assistants monopolized the agenda of many such meetings during the summer and early autumn as well,[62] and a joint meeting of the union leadership and Glav-Textile on October 25 devoted itself almost entirely to this matter.[63] By late 1920, *Tekstil'shchik* suggested that this support was not merely passive or completely opportunistic when it put forward the idea that one-man management actually freed trained personnel for other work and, as responsibility shifted from a seven- to ten-member collegium to one person, that management improved qualitatively.[64] At the end of 1920, some 2,183 of 2,483 factories under *VSNKh* in all industries were included in one-man management,[65] including 1,783 (86 percent) of the large-scale enterprises.[66]

At the same time, the union leadership and Glav-Textile sought to raise productivity through additional steps to improve discipline and to rationalize the use of available resources. As we have seen, instilling the habits of industrial discipline among the Russian textile workers had not previously succeeded on a broad scale. The reports of card playing and sleeping on the job that had punctuated the earliest union communications of late 1917 to early 1918 continued to appear throughout 1920.[67] The workers' comradely courts instituted in 1918 continued to operate[68] but without providing the anticipated improvements in labor discipline. In all of Moscow province, the courts heard only sixty-five cases between June and September 1920,[69] and at the end of the year *Tekstil'shchik* asserted that they were best known for their propensity to exceed their authority. In the words of one union representative: "We know for a fact that many [rank-and-file] members of the union look upon their courts as *officials* fulfilling the role of the former tsarist gendarmes."[70]

The textile industry also participated in the practice of providing voluntary workdays, the *subbotniki*, but with mixed results. Party organizations initiated the first *subbotnik* in July 1919, but few outside the party took part. By January 1920, some factories apparently designated individuals to organize these ostensibly voluntary workdays and passed resolutions affirming their intention to work every Saturday.[71] It would appear, however, that voluntary work without pay did not have as much support as the assertive language of supportive resolutions would suggest, particularly at a time when the workers' desertion of the factories was still a pressing problem. By June 1920, the Union of Textile Workers was openly trying to orchestrate what was supposedly a spontaneous workers' movement.[72] There were, of course, notable successes. As was usual, the Moscow Danilov Factory placed itself in the front rank of support, and hundreds of Moscow textile workers participated in a *subbotnik* in early October devoted to gathering fuel in the forests near the city.[73] The promise of material incentives also drew significant numbers of Moscow *tekstil'shchiki* to a fuel-gathering *subbotnik* in November.[74] Still, reports indicate that such successes do not appear to have been the norm, and on October 1 the union leadership found it necessary to repeat an earlier instruction that *subbotniki* be organized.[75] Had earlier calls been fully heeded, such redundancy clearly would not have been necessary.

The union also restored piecework wages in July 1919–December 1920, one of the targets of its most vehement opposition in the prerevolutionary years. Although not all Russian workers viewed piece rates as an exploitative measure—before World War I metalworkers preferred them and resented their repeal[76]—the textile workers opposed them and had concerned themselves with establishing a *tarif* since 1917. The *tarif* adopted in 1918 rested in principle on linking pay to output. By 1919–1920, however, the high number of rank-and-file complaints of a return to capitalism through the naturalization of wages—encompassing the practice of paying strictly on piece rates—indicates that the practice was strongly reestablished by the end of the civil war. In spite of opposition at the level of the factory workers, the union Labor Department formulated a system of premium pay based on output that the Iaroslavl and Klintsy sections immediately adopted and which the department recommended for the entire industry.[77]

Data are too scant to generalize about the success of piece rates, but individual examples indicate a steady expansion of the system throughout 1920. One provincial union conference, in fact, blamed piece rates for a fall in production in January 1920.[78] The more common reaction, however, was recorded in Moscow province when a union credited the introduction of piece rates with a 65 percent rise in productivity at the Kotov Factory, where each machine produced 17.3 arshins before 1914 but only 5.5 in 1919. Once piecework took effect, output per machine at the Kotov rose to 11.6 arshins, and in the Golutvin Factory production figures even exceeded prewar levels.[79] By March 1920, union leaders had begun to counter criticism of the piecework system by asserting that it was a good idea that had not yet been carried far enough, and its application spread thereafter. Gorshkov told a meeting of factory committee representatives on March 19 that it would be necessary to make premium pay the only wage system in the industry if productivity were to rise, and the delegates resolved to carry out the instructions of the union on this matter completely.[80] In Nizhnii–Novgorod, the union worked out an elaborate system of remuneration that took into account not only the different categories of work, as in the 1918 system, but also different levels of quality within these categories. The Nizhnii–Novgorod system called for bonuses to be paid on an ascending scale of overfulfilled norms: 150 percent of wages paid for 50 percent above the norm, 200 percent pay for 75 percent above the output norm, and so on.[81] In the complex of factors that affected production, it would be impossible to calculate the full impact of this single step. On can say unequivocally, nevertheless, that almost without exception union officials believed piece rates to be a boost to production. On December 17, 1920, N. M. Goncharov of the Iakovlev Cloth Factory in Gomel' attributed the recent rise in output there completely to the use of piece rates. At the same conference, the delegate Pevzner of the Mogilev Factory stated that piece rates had raised productivity in his plant by 50 percent.[82]

Late in 1920, the textile industry also experienced another attempt to solidify the faltering production groups, to utilize supplies in a manner more economically rational, and to raise productivity: shock factories. Ideally, the supply organs would concentrate available raw materials in enterprises designated as shock factories, from which higher output would then be expected. Obviously, this experiment manifests once again an attempt to attack simultaneously a number of related problems of accounting, supervision, organization, and supply with a single measure that concentrated supplies and talented personnel in presently effective industrial enterprises. In June 1920, *VSNKh* designated a group of metal-fabricating plants as shock factories to serve the railroads, and by August 9 a shock group of seventy-nine enterprises from all spheres of the textile industry, employing as many as 40,000 workers, took shape in the Central Industrial Region.[83] At the same time, *Tekstil'shchik* reported enthusiastically about additional textile shock factories in Ivanovo producing results so promising that they should serve as an example for other areas.[84] To supplement the program, *VSNKh* also created twenty-seven so-called model factories on September 9, of which three were in the textile industry.[85]

Union activists below the national level clearly regarded the creation of shock factories to be a response to the failure of the group administrations. On October 3, the Committee of Shock Groups in Ivanovo–Voznesensk singled out the "inert" and "lifeless" character of group administrations as its leading motive for realigning factories.[86] The Union of Textile Workers of Vladimir Province specifically listed inactivity and the fact that existing groups worked only in their own interests as causes when it endorsed the idea of shock factories on October 19.[87] By November, some 67,000 textile workers labored in 208 shock factories in Moscow province alone,[88] and at the end of 1920 there were 408 textile factories among the 1,716 total shock factories in the Soviet Republic.[89]

By definition, however, shock factories could not solve problems on a general scale since the corrective quickly reached a point of diminishing returns. As supplies were concentrated in designated factories, other enterprises suffered even more than previously. In addition, increasing the number of shock factories diluted the overall effectiveness of the program for obvious reasons. Finally, the shock factories themselves did not always work as intended. In some instances, enterprises declined in effectiveness after their designation as a shock factory.[90]

Disappointed Material Aspirations

By this time, the leading supporters of the revolution had certainly learned that central initiatives alone would not produce success and that ventures ultimately depended on the mobilization of local support. In this spirit, Lebedev's lead editorial in the June 1920 issue of *Tekstil'shchik* stressed that the unanimous resolutions of the third national union congress still had not produced any significant improvement in material conditions in the time since they were passed in April. With the exception of a few model examples in Tver and Ivanovo–Voznesensk, he stated, matters had degenerated noticeably even since March.[91]

The tenor of available documents from local areas confirms Lebedev's assessment. When individual factories filed reports at the Nizhnii–Novgorod provincial conference of the union on March 25–26, 1920, they repeated a number of chronic complaints. The Reshetekhin Factory lamented a shortage of fuel and skilled workers, the Molitov stressed that the food crisis had significantly reduced its production, the Trubanov advised that it felt sharply a shortage of qualified specialists, the Gorbatov communicated that a typhus epidemic had reduced its work to half capacity, the Gorbatov and Rastiakin Rope Factories both disclosed that no cultural–educational work was taking place, and the Nizhegorod Wool Factory advised that workers had deserted in large numbers in search of food.[92] During the same month, other union conferences reported again the same problems that had hindered local organization throughout the period of Soviet rule: tension between factory managements and factory committees in Moscow; the need to repeat prior organizational work in Orekhovo–Zuevo; and the necessity to work out the proper scope of factory committee work in Orlov.[93] In August 1920, union reports from Podol'sk did not differ appreciably from the type generally filed a year earlier. The Medvedev and Danilov factories operated, but the Shlikterman had just closed for lack of cotton. The Troitskaia and Dubovits operated partially, but the Strelkov and former Tregubov factories both shut down for want of fuel. Moreover, all reported a shortage of food.[94] Mixed results also arrived from Penza, where different factories reported both increases and decreases in production as well as improvement and deterioration in labor discipline.[95]

Reversing the country's problems proved more difficult than merely resuming work where previous processes had halted. If industrial crops were in short supply, for example, this was due in large measure to the fact that peasants who formerly grew them now used their land, if at all, to raise food. Revitalization thus entailed not only improving the production of industrial crops but resolving the food shortage as a precondition to convincing peasants to plant the desired industrial products. To be specific, peasants in the seventeen flax-producing provinces of central Russia redirected their planting to such a degree that in 1919 the Soviet Republic produced only 5.5 million puds of flax on land that yielded 25 million before the war.[96] When the Soviet Republic lost Turkestan, its principal domestic source of cotton, in 1918, peasants there also shifted to growing food. There was, therefore, no significant supply of cotton on hand for the Central Industrial Region when the Reds recaptured the area in 1919.[97] Cotton production for the year fell to 750,000 puds, or 4.7 percent of the total national demand for cotton. In the first quarter of 1920, reported cotton production stood at 130,000 puds, or 3.25 percent of demand. Rumors nevertheless circulated that over 8 million puds were on hand in Turkestan. Even if true, there was not transportation available for so much as a small fraction of that figure.[98] In addition, hemp output fell in 1919 to 6 million puds, in contrast to a prewar total of 20 million, and wool production contracted to 100,000 puds, compared with 6 million before the war.[99]

In addition, workers continued to convert factory land to food production. According to one count, 82 enterprises used 11,432 desiatins of factory land for agriculture in 1919, a total that grew to 131 enterprises and 16,727 desiatins in 1920.[100] In line with this trend of devoting attention to the acquisition of food, the Podol'sk Section of the union did not particularly distinguish itself in factory agriculture but channeled significant energy in August 1919 toward preparing a provisioning expedition to Omsk. Virtually nothing appears in the records of the section for that month that would indicate it gave any attention to textile production.[101]

Factory closings actually increased as the civil war wound down. Although some regions such as Serpukhov reported approximately half of their factories open[102] and in Podol'sk reopenings outnumbered closings,[103] Nogin estimated in September 1919 that only 10 percent of the total number of nationalized textile factories were in operation.[104] By March 1, 1920, only 30 of 130 textile enterprises, employing 30,000 in place of the former 200,000, remained open in the "Russian Manchester," Ivanovo–Voznesensk. On May 1, these figures fell to eight factories and 5,000 workers.[105] *Tekstil'shchik* reported only 7.1 percent of the country's spindles and 15.1 percent of the looms in operation by mid-1920.[106] In addition to the havoc caused by the food shortages, army mobilizations continued to drain away scarce skilled workers, who tended to be male in disproportionate numbers. In one striking example that illustrates the special problems of the textile industry in this regard, a single Petrograd enterprise reported in October 1919 that it employed an insignificant number of male workers. When the army took even these few men, however, the plant closed, because these were the skilled workers who operated the plant's steam equipment.[107]

Renewed expectations led to disappointments in the area of supply as well. In late September 1919, Nogin told the congress of representatives of the group administrations and Gub-Textiles that future supply would improve despite present problems.[108] When he reviewed the supply situation in February 1921, however, he reported that no discernable improvement had occurred before the final quarter of 1920.[109] Indeed, when Guseva of the Nizhnii–Novgorod Province Union of Textile Workers met with the representatives of two individual factories, two group administrations, and fourteen workers' *arteli* in October 1920, the local delegates, as before, overwhelmingly focused their remarks on the familiar shortages of food, raw materials, and skilled workers.[110] From the national perspective, the fact that Glav-Textile met jointly with the Commissariats of Food and Land from October 7 through 12 to plot a supply strategy for the textile industry indicates the determination of central organs to rectify existing shortfalls. It also shows, however, that even near the end of 1920 planning and coordination had not passed the stage of general discussion.[111]

Consequently, national and local reports continued to document with regularity conditions for the workers even more deplorable than those described earlier in the civil war. Nogin conceded that "it is well known to all of us that in very few factories are living conditions fit for human habitation."[112] The delegate Zaks reported to the Third Congress of the Union of Textile Workers on April 20 on the need for more hygienic conditions throughout the industry,[113] and "Rabotnitsa" described graphically in *Tesktil'shchik* how the children of textile workers regularly died of cold and hunger.[114] Local communications were even more specific. At the end of 1919, typhus hit the Podol'sk Tregurov Silk Factory so hard that even small children were forced to work if the plant were to remain open. The level of animosity between Tregurov workers and union officials was such that initially some workers actually resisted the union's efforts to combat the disease, and nothing had been accomplished in the area of proposed maternity care.[115] Rapporteurs from the Voznesensk Textile Mill considered living conditions even worse than when the Knoop family owned the enterprise despite the fact that fewer workers now occupied the available space.[116] In a separate instance, Kutuzov of Glav-Textile seconded the assessment that living conditions were better before the revolution: two families at the Likino Mill in Orekhovo–Zuevo lived in space occupied by one when the Morozov family owned the plant; in the barracks, five now lived where three had under former proprietors.[117] A communication from Iaroslavl in June 1920 compounded the pessimism by advising that most food excursions organized by the workers ended without success.[118] On a separate issue, the eighty textile plants surveyed in the Central Industrial Region reported that underage workers still constituted 6.7 percent of the labor force (slightly under the national average of 8.2 percent) despite corrective legislation.[119]

No significant improvement took place in the second half of 1920. An assessment of the condition of flax workers made in August concluded that a national decline had occurred. No one had repaired living quarters since before the revolution, and the poor ventilation, dusty air, lack of temperature control, and general sanitary laxness in the factories made bronchitis a common malady.[120] Twenty families in Simbirsk shared a single stove, and working mothers had no recourse but to leave small children in the care of those too old to work. Of 650 children at the textile factories there, none was in a nursery. Not one of 850 invalids was receiving the special care so frequently mentioned in public resolutions.[121] On October 28, Kutuzov again went on record in a meeting with the factory committee of the Veshetikhinsk Knitting Mill as stressing the need to improve living conditions, especially the remodeling of dormitories. The redundant nature of the declaration, however, made clear how seldom similar earlier resolutions had been carried out.[122] Also in late October, textile workers in Vladimir province complained that their conditions were inferior to those of their counterparts in water and rail transport.[123]

Among the most hard-pressed was the industry's majority, its female workers, who exercised scant influence over their trade union and failed to penetrate leadership positions in appreciable numbers. This held true in union organizations of various scales. As of September 1919, the three officials who led the 1,080 men and 1,463 women of the Nizhnii–Novgorod Section of the union were all men. In the same month, women constituted 2,597 of the 4,588 members of the Podol'sk Section but held none of the five union offices. In October 1919, women outnumbered men 91 to 84 in the Viatka branch but occupied only one of four leadership posts. At the same moment in the Nefoekhta branch, women constituted 5,581 of the 8,488 members, but they held only three of eight responsible posts. In Moscow, men held eight of nine union offices despite being outnumbered 22,000 to 18,000 in the general membership.[124] On December 23, 1919, chairman of the Podol'sk Section of the union, A. Riabov, called special attention to the lack of definitive organizational work among women and proposed both a general district conference of all women workers and a separate conference of female textile workers.[125] Such isolated local actions notwithstanding, only 24 of 358 delegates in attendance were women when the Third Congress of the Union of Textile Workers opened the following April.[126] By December 1920, only 12 of 38 union sections included women in leadership posts. The total of women involved numbered 16, of whom 9 were Communists. To employ a different measure, women constituted 57.9 percent of the union membership at the end of the civil war but only 8.2 percent of the leadership at the section level or above.[127]

The revolution and three years of shared suffering apparently failed to alter the low regard in which male textile workers held the female majority in the factories. In June 1920, an open letter from the union leaders to the British textile workers complained in *Tesktil'shchik*: "The union is constantly taking into account the low level of consciousness of the female workers."[128] In October, "Rabotnitsa" reported an incident that suggested male textile workers shared the opinion of their union leadership on this issue. When the Zubov Factory in Klintsy temporarily closed, the group administration decided to transfer a portion of the female weavers to the Stodol' Factory. Male weavers at the Stodol', in what the author described as "a burst of masculine pride," labeled the move an attempt to provide busywork for the women and threatened to stop work as soon as the women began. Although the administration did not succumb to the ultimatum, "Rabotnitsa" concluded that the point of the incident was that women were fighting the same battles repeatedly.[129] At the end of the civil war, returning soldiers would commonly demand that they be given jobs presently held by females.

Amid such attitudes, women dominated the rolls of the unemployed. The Nefoekhta union reported in October 1919 that women accounted for 1,416 of the 1,456 unemployed there.[130] On September 1, 1920, 44.2 percent of the textile workers in Ivanovo-Voznesensk were women, but they accounted for 96.2 percent of the unemployed. In Kostroma, the figures were 32 percent and 94.5 percent, respectively. The monthly statistics provided by the Pavlovskii–Posad Section between January and October 1920 on the representation of women in the factories fluctuated between a high of 67.6 percent in May to a low of 56 percent in September. In the same period, however, women made up from 92 percent of the unemployed in January to 100 percent in July and October. Nationally, union figures reported that women made up 58.4 percent of the total work force in the industry at the time and averaged 93 percent of the unemployed. As the analyst who reported these figures for the union journal noted, attitudes toward women, however important, did not provide a full explanation of the differences in unemployment figures according to gender. Also important were the facts that the Red Army rapidly mobilized unemployed males and that in the textile industry males were more highly represented among the skilled workers, enabling them to find work more easily during this period when skilled labor was in demand.[131] These mitigating factors, though noteworthy, by no means explain the full degree of the disparity.

The Dilemma of Local Officials

If the task of organization presented a bleak picture when viewed from the center, it caused even greater consternation among those charged with its implementation at lower administrative levels. Union activists in the field and local officials continued to protect themselves from encroachments from above, on the one hand, while simultaneously petitioning for greater support from superordinate bodies, on the other. Underlying this conundrum, in part, was the weak leadership provided by responsible organs. In June

1920, the *VSNKh* presidium—despite the status of the textile industry as a "model" for other nationalized spheres of production—
deemed "the Glav-Textile problem" so serious that it sent Rudzutak
to examine the situation personally.[132] In November, the Central
Committee of the Union of Textile workers also scrutinized Glav-
Textile and, finding its performance substandard, decided to send
an additional five representatives to strengthen its apparatus.[133]
Organs subordinate to Glav-Textile, to be sure, lacked this kind of
recourse and had to deal with the institution as they found it. The
kust of Egor'ev-Ramen Cotton Factories, for example, retreated in
1920 from its uncompromising tone of the previous year when
addressing Glav-Textile but with little improvement in the results.
"The purpose of our report is to appeal to the *Glavnoe pravlenie* for
help," began one of its communications that suggested correctives
to sources of ongoing frustration: that Glav-Textile not take over
factories without the prior knowledge of the group administration
involved; that the center send instructions to group administrations
before the date by which assigned tasks were to be completed
instead of after, as was often presently the case; and that Glav-
Textile regularly publish questions raised by group administrations
and its responses.[134] In a different instance, the request of the
Karachev (Orlov province) Section for an instructor from the
union's national Organizational–Instructional Department was refused because the central organ had none to send.[135] In August
1920, *Tesktil'shchik* carried reports of places "where union representatives do not have the opportunity to go for weeks and even
months."[136]

The frustration intermediate officials felt from a lack of support
from above was compounded by the animosities and suspicions
encountered within their own jurisdictions. In the second half of
1919, the Naro–Fominsk Union of Textile Workers reported to the
Kaluga Provincial Council of the National Economy that many of
the textile workers and employees of the six large-scale, mechanized
factories and approximately twenty smaller enterprises in the province were not aware that a national textile union existed. Those
workers willing to organize wanted to do so only on the scale of
their own factory and resisted subordination to higher bodies.[137]
Union communications from Bogorodsk advised that fewer than
half the workers would attend meetings if union instructors were

scheduled to discuss questions of organization,[138] a statistic corroborated by a central Organizational Department report that cited only 40 percent turnout at sessions it held at fifteen Bogorodsk factories.[139]

Even at the center of the industry, where the Moscow Provincial Union of Textile Workers encompassed ten sections, 449 factories, and 194,562 workers and employees in May 1920, organizational work frequently miscarried. When regional representatives of the Moscow union met on May 14, the provincial chairman, Mukhin, disclosed that the provincial apparatus had experienced scant success even in organizing its own activities, despite devoting virtually all its time to the task. He advised that there were no internal union ties within the province, an assessment that Limanov, who spoke for the Organizational Department, echoed. In some areas, Limanov lamented, workers' meetings had not been held in more than six months, and uncoordinated *raion* unions were not yet linked to the provincial union. Although Iasenev also added that each union section still acted independently and the delegate Tarasov blamed the provincial conference held February 5 for not resolving the question of organization correctly, this meeting adopted only modest proposals: making a list of workers with administrative experience and vowing to improve communications between the provincial apparatus and the outlying areas.[140] There were, of course, also reports of local successes that reached the center, where they especially raised hopes of better coordination of supply.[141] As the delegate Zhavoronkov told a conference of group administrations in November 1919, however, the general organization of the *kusty* had been done "mechanically" and without achieving broader aims.[142]

To emphasize this dimension of the situation contradicts those who believe that the Soviet government had achieved a high degree of organization and centralization during the civil war and that by 1920 it had at least set in motion an extensive militarization of labor. In actual fact, *Sovnarkom* did create a commission to work out the militarization of labor on December 19, 1919, which Mikhail Tomsky, chairman of the Central Council of Trade Unions, communicated to the Central Committee of the Union of Textile Workers early in January 1920.[143] Such a militarization of labor could not have amounted to more than wishful thinking in the textile industry at this time. In addition to the material already presented that supports this assertion, one should consider another piece of evidence. Glav-Textile did not hear its preliminary report on this project, "On the Militarization of the Textile Industry," until November 3, 1920. Even then, the action taken amounted to no more than instructing the Administrative Department to cooperate with existing production commissions to work out a detailed plan for militarizing the entire industry in the future.[144]

Rather than heading a labor force fashioned on a military model, then, Glav-Textile and the Union of Textile Workers expended a great deal of energy just trying to keep their forces from dispersing completely. Despite precautions taken, a large number of textile factories could not reopen after the summer break of 1919 because a sufficient complement of workers failed to return.[145] Between then and the autumn of 1920, the union suffered an acute loss of skilled workers to the more remunerative metal and railroad industries. One local report published in *Tesktil'shchik* declared that only "women and invalids" remained behind.[146] In Bol'shevo, even three members of the factory management fled in November 1920 for what were euphemistically described as "various reasons," and the Organizational Department of the local union could not function because it had lost its key personnel.[147] The union had certainly addressed the flight of workers before, but from the middle of 1919 it began to treat it as a discipline problem in which it might enlist the aid of the political police and the Military Commission. Short of this, the union enlisted local officials with experience in the disciplinary courts to operate newly formed subdepartments to combat the problem. Finally, the leadership decided to use drama, concerts, plays, and musical evenings to propagandize against desertion and to employ satiric posters depicting the factory owners, tsarist generals, the nobility, and private traders as the direct beneficiaries of such acts.[148]

Education and propaganda in the industry continued to yield but mixed results through the end of 1920. New optimism in this area emerged at the third union congress, even though its resolutions continued to repeat well-established goals: eradication of illiteracy; intensification of political lectures; fostering of a broader understanding of economics; short-term technical courses; and establishing a close working relationship with the People's Commissariat of Education.[149] The union section of Orekhovo–Zuevo, already distinguished for vigorous educational efforts, reported what it considered measurable results in its report for the final quarter of 1919. The local Cultural–Educational Department created a small library in the factory formerly owned by Vikula Morozov by centralizing available books and magazines. Three technical courses attracted only eight students, but eighty-six attended evening classes, very likely for basic literacy instruction. In addition, a member of the union's Engineering Department established the Orekhovo Technical School.[150] The union in Kineshma reported that fifteen factories had established five kindergartens, six nurseries, three workers' clubs, six theaters where 120 performances took place, six literacy schools, and one technical course.[151] On February 5, 1920, Ammosova, head of the national Cultural–Educational Department of the union, announced at a Moscow provincial conference that almost all sections of the union had undertaken some form of technical instruction, and a technical school had opened in Moscow itself.[152]

Negative reports proved more dominant. Contradicting Ammosova's optimistic assessment of the situation in Moscow province, the Moscow Section of the union officially noted that a shortage of qualified instructors had greatly inhibited its cultural–educational work in the July 1, 1919–February 1, 1920 period. The union had opened one of two planned libraries, but reported no other achievements.[153] In September, the Moscow provincial union communicated that its cultural work since February consisted only of distributing two journals, *Tekstil'shchik* and *Professional'noe dvizhenie*, in the factories.[154] From Podol'sk, the union secretary, M. Leonov, listed even the establishment of literacy schools as a future objective at the end of 1919.[155] Moreover, attrition plagued those educational programs in existence. Only 18 of 31 students in Iartsev completed a technical course begun July 16, 1919; 12 of 18 dropped out of a similar course, held August 1, 1919–April 20, 1920, at the Voznesensk Factory in Naro-Fominsk; by January 1920, only 40 remained in a course in the Serpukov Factory that 100 had begun on October 1, 1919.[156]

In addition to reiterating that good intentions expressed in resolutions did not always bear the desired results, communications on education also made clear another point more troubling to committed activists both in the center and at other levels: the workers were frequently not interested in the messages that did reach them. In the final year and one-half of the civil war, mounting evidence began to call into serious question the widespread assumption that national disruptions and a shortage of activists alone explained the lack of local cooperation from the workers. As N. M. Goncharov reported from the Iakovlev Cloth Factory in Gomel' at the end of 1920, cultural–educational work often amounted to little more than directing the workers' social activities. His union staged two sparsely attended dramatic performances per week, and workers absolutely never attended lectures or meetings. On the other hand, Goncharov noted, although "any interest in cultural work is absent among the working masses . . . the evening dances attract a significant number of workers."[157]

Data that Ammosova compiled nationally at the end of 1920 illustrate that the situation in Gomel' reflected a broader trend of entertaining workers as much as instructing them. The Ivanovo–Voznesensk Section organized 518 meetings and 427 lectures during the year, but also presented 92 concerts, 438 dramatic performances, and 273 films. In Simbirsk, the union held 118 meetings and 36 lectures, while also listing 35 concerts, 184 dramas, and 40 films. Serpukhov reported 162 meetings and 64 lectures as well as 40 concerts, 220 dramatic performances, and 162 films. Instruction and entertainment were not always distinct, of course, and data were frequently sketchy and incomplete, but we can discern a trend toward a heavy emphasis on entertainment. In all, union sections held 1,823 meetings and 880 lectures in 1920 while presenting 432 concerts, 1,980 dramatic performances, and 1,082 films.[158]

Communications in this period also began to make more frequent and specific reference to the link between educational work and cadre formation. When Korolev addressed the Ninth Provincial Congress of the Ivanovo–Voznesensk Union of Textile Workers, held May 6–8, 1920, his prescriptions were remarkably free of the platitudes that had characterized such speeches in 1917–1918. His focus was immediate and pragmatic: without expanding technical and professional training, there was no hope of alleviating the shortages of qualified personnel in practically all spheres of responsibility.[159] Ammosova articulated a similar sentiment when she called for professional education to take root in localities other than Moscow,[160] as did the delegate Mokovskii in a report on cadre shortages in Gomel'.[161]

Conclusion

While we can discern a great deal of continuity between the problems of the second half of 1919 through the introduction of NEP with those of the earlier period of Soviet rule, the period illustrates how rising expectations exacerbated tasks and goals. The mutual isolation of institutions actually deepened at this time. This not only caused continuing concern over organizational questions, but—more to the point—demonstrated that existing problems were not the by-products of the war emergency alone or of its accompanying focus in public pronouncements, bourgeois sabotage. By 1919–1920, officials speaking in the name of the working class had established the supervisory authority over the administrative bureaucracy called for in prerevolutionary formulas for the transition to socialism. Instead of creating harmony, the gaps among responsible institutions—and between such institutions and their respective constituencies—widened as the threat of military defeat receded.

In the final year and a half of the civil war, experience impressed much more strongly upon the central officials of the union and Glav-Textile and those at other levels the degree to which they were functioning largely independently. The group administrations created so successfully in the first half of 1919 failed to provide the desired impetus for coordinating the industry on the scale of the Soviet Republic. When group administrations worked energetically, which was far from the universal case, they did so chiefly in the immediate interests of their own *kust*. Even within successful production groups, internal divisions and rivalries deepened. The prospect of victory brought with it higher expectations, especially for an improved quality of life, but did nothing to alter institutional behavior patterns.

Neither the end of the war nor the introduction of NEP immediately altered the rhythm of the textile industry. As in other spheres of manufacturing, there were attempts increasingly to tighten discipline right up to the Tenth Party Congress. In the textile industry, however, these met with no greater success than had earlier overtures. In early 1921, resistance from the rank and file became, if anything, even more strident and impatience with the repetitive rhetoric of national and local officials grew. As we have seen, the production groups continued to fall so far short of expectations that the union disbanded them in February 1921, just weeks before the pivotal Tenth Party Congress. Discipline continued to fall even after NEP began, and indeed by the autumn of 1921 criticism of the union leadership from rank-and-file union representatives reached new levels of rancor. Union chairman Kutuzov admitted the failure of local administration, yet the union leadership and Glav-Textile continued to be targets of criticism, particularly for a perceived indifference toward the local areas.

NEP failed to alleviate factory conditions in the short term. If anything, the end of the fighting exacerbated unemployment and internal frictions within the work force. Returning soldiers demanded that employed females be dismissed and their jobs given to military veterans, often on the spot. Still, production recovered too slowly in the immediate postwar years to employ a significant mass of those out of work or to introduce a meaningful influx of manufactured textile goods into the economy. Documents of 1921, in particular, show little change on major issues from the previous three years.[162] What did change was the apparent ability of the union leadership to ameliorate the tone of public, if not internal, union discourse. This was especially pronounced in the content of union publications. *Tekstil'shchik*, in particular, abruptly replaced the open criticism and exchange that characterized the journal during the civil war years with more muted tones in 1921. In the end, instituting normalcy would strain Bolshevik ingenuity even more directly than had the civil war.

6

Divergent Perceptions of Revolution: Working-Class Politics in a Workers' State

By the end of 1920, the leaders of the Bolshevik Revolution found themselves not only widely separated from their original base of support but also presiding over institutions and constituencies far from mutually integrated. During 1917–1920, while the party leadership pursued the progression of policies that preceded the introduction of NEP, activists and mass workers at all noncentral levels were far more influenced by local demands and allegiances than by the long-term concerns of the revolutionary elite. In the localities and factories, economic complaints and retribution against the beneficiaries of the prerevolutionary order consumed the energy that revolutionary ideological pronouncements and slogans had reserved for the establishment of effective working-class supervision over national administration. The style of administration prevalent in the early years of Soviet rule, therefore, was a process that further *separated* the central authorities from society, not the reverse. Centralization as instituted in 1918–1920 did not solve problems but rather postponed the moment of reckoning.

In the textile industry, the local component of this situation was not merely the result of widespread political immaturity or a by-product of low ideological sophistication. While both existed widely, to make them the primary focus of interpretation imposes categories of assessment on the unskilled and semiskilled that are better suited to evaluating the political behavior of (and indeed were employed by) conscious workers and the revolutionary elite. If viewed in terms of agendas they set for themselves, however, the local rank and file in the textile industry turned out to be remarkably consistent in their motivations and cognizant of their objectives. Documentation for the present study consists largely of the records and communications of working-class institutions. Mass workers, by definition, did not prepare such documents directly. Their aspirations enter the historical record through reports by their working-class representatives of the actions mass workers took and of the pressures the mass brought to bear at all levels, including in the factory. Judging from the issues on which officials and institutions felt relentlessly pressured, local interests continued to be identified in 1918–1920 much as they had been against the Imperial and Provisional governments in 1917. The high degree of noncooperation with initiatives from above after October 1917, therefore, was grounded not simply in a lack of political consciousness but in the proclivities that underlay the defense of personal and factory interests—the very proclivities that were manifested initially in ideas such as workers' control. In one sense, although the unskilled continued by trial and error only to approach the level of experience and consciousness obtained by the workers' "vanguard" before October 1917, the ability to sustain the pursuit of their own agendas demonstrates greater constancy than that for which both local textile officials and workers are usually credited.

Nevertheless, situations such as those found in the textile industry worked to the detriment of establishing administrative cooperation on a national scale. We have seen how local actors repeatedly proved unwilling to surrender to the central officials' centralizing proclivities in the absence of at least the prospect of tangible material gain. In practice, local officials' tolerance of central authority derived as much from a desire for aid and moral support as from revolutionary enthusiasm. As for the rank and file, we have encountered significant evidence of both their ideological indifference and their acute sense of self-preservation, propensities that ensured that their local or even factory representatives did not always speak precisely in the name of their constituencies. Yet local contumacy in this context never crystallized into a meaningful attempt to overthrow the revolution from within. The resultant relationships between the center and the locales were thus an uneasy compromise between unbridled localism, at one extreme, and the ideals projected in national state and party policy statements and declarations, at the other. By the end of 1920, insofar as local actors in the textile industry marginally displayed more of the characteristics generally associated with political consciousness—an awareness of the common interests of industrial workers, the capacity for sustained organization, a recognition of political and socioeconomic linkage, and an appreciation of long-term improvements over short-term gratification—they did so as frequently against the policies of the government that spoke in the name of the working class as in their support.

Material and social preconditions in the textile industry figure strongly in explaining the pattern of revolutionary development in 1917–1920. Certainly, the prerevolutionary circumstances of life and work fostered in the textile labor force an overriding desire to strike out at those whom they identified as responsible for their plight. The legacy of submissiveness of the female majority in the industry, their systematic exclusion from labor politics, the overwhelming numerical preponderance of unskilled, politically less conscious laborers, a high incidence of geographic isolation, and the absence of a strong tradition of sustained organization even within the circumscribed limits possible under tsarism all influenced the volatile but ephemeral character of labor activity in the industry by 1917. Lacking the level of conscious political response in evidence among metal and railroad workers, the textile workers were strongly predisposed before February 1917 toward the fixation on economic betterment that they exhibited thereafter.

But this is not a case to be explained by socioeconomic determinism, and the actions of the textile labor force also showed a developing—albeit often indirectly and even impressionistically articulated—sense of a range of politically palatable options. Deserving special emphasis, therefore, is the fact that officials and workers in the textile industry, despite the strongly economic orientation of emerging demands, did not unreservedly support alternatives promising the most expedient economic solutions. Officials and the rank and file clearly demonstrated one boundary of their minimum revolutionary expectations when social antagonisms repeatedly motivated them to eschew the needed expertise of the prerevolutionary management, even when the alternative was total self-reliance. Thus, even though the issue of survival largely dominated immediate objectives, it was not to be achieved at the price of reinstituting the old order in the form of either White generals or factory proprietors. Neither in 1917 nor thereafter did the textile workers rally either to the highest bidder or to the most clever demagogue but gave their support, if at all, to those whose immediate actions most closely matched their own self-defined aspirations.

The Early Soviet Experience in Perspective

There is much to suggest that the story of the textile industry reflects a highly representative experience in the early history of Soviet Russia. Since the revolutionary urban "vanguard" of 1917 dispersed from the factories by mid-1918, the example of an industry dominated by the unskilled who relied largely on their own devices for survival could indeed be pertinent far beyond the geographic and industrial foci of this work. Evidentiary corroboration of this point must, of course, await the appearance of parallel studies. What can be stated unequivocally is that the textile industry experienced in microcosm a large number of the conflicts that became nationally important by 1920–1921. At the highest levels of the Bolshevik party, the Workers' Opposition, led by former workers like Alexander Shliapnikov, represented the interests of those who considered themselves loyal Bolsheviks but who were hostile to the emerging bureaucracy. In retaining much of the spirit of 1917, the Workers' Opposition demanded greater intraparty democracy, flexibility for local party organizations and trade unions in the face of national party authority, and an ongoing commitment to the ideological expectations of the prerevolutionary party. Regarding the final point, its members were especially adamant in their distrust of the use of bourgeois specialists and resentful of the privileges accorded them in a workers' state. These were, as we have seen, among the concerns that preoccupied those trying to organize and revitalize textile manufacturing.[1] In short, the issues that dominated textile politics throughout 1917–1921 were not confined to this single industry but resonated in the battles at the highest reaches of the Bolshevik party.

In addition, the concerns that surfaced in the textile industry offer insight into the emergence of authoritarian proclivities in the

early Soviet system. Indeed, those who seek the beginnings of
Soviet authoritarianism in 1918-1920 are certainly correct[2] as long
as one makes an important distinction: the difference between the
centralization of decision making (which took place) and the establishment of extensive hegemony of central institutions (which did
not). We must therefore bear in mind that although the Bolsheviks
were ideologically predisposed toward building a centralized,
planned system[3] and they relied increasingly on measures of greater
centralization as they faced the crises of 1918-1920, the explanation
of the authoritarianism that emerged in 1918-1920 cannot rest with
ideology alone. Bolshevik prerevolutionary guidelines for the transition to socialism equally emphasized that mass mobilization
would be instrumental in establishing centralized planning, production, and distribution—not as ends in themselves but as measures of
general economic rationalization and public supervision. Therefore, central direction was not alien to prerevolutionary Bolshevik
ideology, but the Bolsheviks also did not institute it gratuitously
after 1917. On the contrary, centralization increased in response to
the nonmaterialization of local and regional cooperation. In this
sense, the style of centralization that actually emerged cannot be
viewed other than as an important failure of Bolshevik efforts at
mass mobilization. It also reflected a growing suspicion and animosity toward local officials by those serving at the top of the
apparatus, a phenomenon that would continue throughout NEP
and surface significantly during the First Five-Year Plan.

The issue that frequently exacerbated animosity among the constituency of the revolution was the status of the bourgeois specialists. No one would seriously deny that class antagonisms plagued
institutions from the factory level through the central People's
Commissariats. In the textile industry, evidence of this is offered by
the battles within Centro-Textile, over the composition of group
administrations, and regarding one-man management. The repetitive directives reserving one-third of administrative posts for professional experts furnish further corroboration, as do the emotional
and disillusioned outbursts at conferences and in communications
about the continued influence of former owners and managers in
the industry in the years that followed nationalization. As we have
seen, the typical confrontation on this issue came in the form of a

superordinate official justifying the continued reliance on specialists against strong sentiments from below for their ouster.

Conflicts of expectations such as those over the bourgeois specialists illustrate the high degree of fragmentation among those who considered themselves supporters of the revolution. In other words, not all those who supported the "revolution" expected the same things from it, and this diversity of expectations certainly manifested itself in institutional relationships. Hence, although documents of the period habitually express themselves in terms of distinctions between the "center" and the "locales," these terms are more suggestive than descriptive. What existed was not a bilateral relationship but one encompassing a complex of vertical and horizontal dynamics. Neither the "center" nor "locales" were static phenomena. The national leadership of a trade union, for example, may have appeared as part of the "center" from the perspective of an outlying town or factory, but national union officials and the country's political leaders by no means viewed themselves as possessing equivalent authority. Indeed, the attempt by the highest union leadership to protect its independence of action from the encroachments of supreme party and state organs (while simultaneously extending its dominion over intermediate and local union organizations) was a recurrent theme of 1917–1920. The existence of provincial and *oblast'* organizations within each institutional hierarchy interjected another administrative layer that precluded easy demarcation between central and local identifications. "Local," for its part, could denote the lowest level of organized administration, factory politics, or even the general mood of a given locale. In short, the distinctions between the center and the local areas were real, and they were so understood by contemporaries. At the same time, this boundary was above all a perceived and fluid orientation rather than a precise functional description. Except for the very highest party and state bodies, which were consistently "central," an insitution that appeared as part of the assertive "center" in one instance might adopt a defensive or parochial "local" posture in different circumstances.

Divergent definitions of the "revolution" also shaped the character of politics. In the national arena, questions of ideology, legitimacy, and party allegiance were points of serious contention, but the influence of politics in this sense declines as one moves from higher to increasingly lower administrative levels.[4] It is undeniable that the issue of party affiliation carried even greater significance after October 1917 than before among both Bolsheviks and their rivals, but it is equally true that in 1918–1920 the character of Bolshevik party consciousness changed fundamentally. The rapid expansion of the ruling party radically altered party composition and redefined the meaning of party membership. In conjunction with these factors, the low level of local party development inhibited the extension of the concerns of the Bolshevik leadership into the local areas. But this is not strictly a Bolshevik issue. The agendas of all parties of the revolutionary elite shared the concerns, if not the views, of the leading Bolsheviks, but these were not, as we have seen, the issues that shaped local politics. Local textile workers, as our evidence has shown, did not articulate their wants in terms of the conflicting programs of the revolutionary intelligentsia.

What, then, prevented the Bolsheviks from losing power as a result of their failure to satisfy local needs? While this question cannot be answered with finality, the explanation lies above all in the fact that the Bolsheviks became publicly identified in 1917 as the party most likely to introduce sweeping changes. They did not lose this reputation among textile workers in 1918–1920 even though Bolshevik authority was challenged nationally in the soviets, unions, and so on through 1920 and beyond. How can this be? The answer lies in the fact that one must look beyond such organs for a full explanation of early Soviet politics because even officials of working-class institutions did not speak directly for local constituencies. As we have seen, duly elected factory representatives clashed regularly with the rank and file and reflected the local mood only in approximate ways. In other words, the factory workers clearly and repeatedly made known their expectations, but in the process the textile industry produced no significant evidence of strong local sentiment for the actual replacement of the Bolsheviks. The government and its policies received criticism, to be sure, and under prevailing conditions this could only be expected. Criticism, however, did not escalate into a widely expressed desire at the local level for the removal of the Bolsheviks. On the contrary, local textile workers consistently put forward agendas overhwelmingly expressed in terms of obtaining more of what was expected from the Bolshevik Revolution, not of removing the Bolsheviks in favor of another alternative. Albeit riddled with problems, the party's mobilization and propaganda efforts in combination with its pattern of mollifying local problems on an ad hoc basis proved sufficient to prevent discontent from crystallizing into a truly viable movement for the Bolsheviks' ouster.

In the end, attempting to resolve the tension among the multiple central and local expectations of the revolution was a task that would dominate not only 1917–1920 but also the first decade of Soviet rule and, in a broad sense, the Soviet agenda to this day. After October 1917, the revolutionary regime attempted to institute a new order without having resolved the problems of the old. Similarly, the introduction of NEP provided a new direction without having satisfied the original impulses behind mass revolutionary support, especially in the area of creating a more egalitarian socioeconomic order. This resulted in widespread resentment of the profiteering nepmen, disillusionment among the revolutionary faithful at conspicuous social decadence in the 1920s and at the opportunities available to the former bougeoisie in business, resentment of the technical intelligentsia by workers in industry, and the heated struggles that occurred in every sphere of artistic and intellectual endeavor.

The slow recovery of Soviet industry in the 1920s complicated the search for solutions, and the need for consumer goods such as textiles was felt especially acutely. The production of cotton fabrics began to approach 1913 levels only in 1926, unemployment proliferated, and the proposed formation of a national textile trust failed to bring forward impressive results. The civil war also had revitalized *kustar'*, an impulse that found a congenial atmosphere in the conditions of NEP. The recurrent complaints about the dearth of manufactured goods, however, indicate that both sources of textile goods failed to produce at levels commensurate with national demand.

Final Observations

In an enduring sense, the attempt to transform the textile industry of Soviet Russia in 1917–1920 *is* "the birth of Soviet industry." The parameters of this reorientation were broader than those of individual cities, and the prevailing conditions suggest a broad relevance of the textile experience, especially in light of party leaders' propensity to cite the industry as a model during nationalization. In this case study, we see that much of the consolidation of Soviet power was, like the revolution itself, dependent on support from the bottom of society upward. Although hard-pressed, even mass workers were not simply victims of events in 1918–1920. They were overwhelmingly influenced by economic goals and localism, both of which the revolutionary elite attributed to low political consciousness. But the textile workers proved neither defenseless nor the pawns of manipulation and demagoguery. It was not only political immaturity that restrained their support of Bolshevik policies after October 1917, unless one denies them the right to define their own aims. Their insistent calls for economic betterment, the redress of inequities, and the prosecution of social antagonisms may have fallen short of the leading Bolsheviks' ideological prescriptions for a socialist revolution, but in light of comparative history they are common and fully legitimate revolutionary aspirations. As the officials and workers of the textile industry emerged from the crisis of 1917–1920, therefore, they did so with their own perception of the revolution as a remedy of former conditions, even if this contradicted aspirations articulated by the elite that spoke in their name.

Notes

Chapter 1

1. The essentials of this program are found in "Can the Bolsheviks Retain State Power?" See Vladimir I. Lenin, *Polnoe sobranie sochineniia*, 5th ed. (hereafter *PSS*) (Moscow: Gosudarstvennoe izdatel'stvo politicheskoi literatury, 1962), vol. 34, 305–06, 309, 311–12, 320.

2. The Central Industrial Region consisted of Moscow, Vladimir, Kostroma, Tver, Kaluga, Iaroslavl, Nizhnii–Novgorod, Tula, and Riazan provinces. For organizational purposes, the Moscow *Oblast'* Bureau of the Bolshevik party included Voronezh, Orlov, Smolensk, and Tambov provinces.

3. Alexander Rabinowitch, *The Bolsheviks Come to Power: The Revolution of 1917 in Petrograd* (New York: Norton, 1976); Steven A. Smith, *Red Petrograd: Revolution in the Factories, 1917–1918* (Cambridge: Cambridge University Press, 1983); idem, "Craft Consciousness, Class Consciousness: Petrograd 1917," *History Workshop* 11 (Spring 1981): 33–56; David Mandel, *The Petrograd Workers and the Fall of the Old Regime: From the February Revolution to the July Days, 1917* (New York: St. Martin's, 1983); idem, *The Petrograd Workers and the Soviet Seizure of Power: From the July Days to July 1918* (New York: St. Martin's, 1984); Diane Koenker, *Moscow Workers and the 1917 Revolution* (Princeton, N.J.: Princeton University Press, 1981); William G. Rosenberg, "The Democratization of Russia's Railroads in 1917," *American Historical Review* 86 (1981): 983–1008; William G. Rosenberg and Diane Koenker, "Skilled Workers and the Strike Movement in Revolutionary Russia," *Journal of Social History* (Summer 1986): 605–29; idem, "The Limits of Formal Protest: Worker Activism and Social Polarization in Petrograd and Moscow, March to October, 1917," *American Historical Review* 92 (1987): 296–326; Tsuyoshi Hasegawa, *The February Revolution: Petrograd, 1917* (Seattle: University of Washington Press, 1981); Tim McDaniel, *Autocracy, Capitalism, and Revolution in Russia* (Berkeley and Los Angeles: University of California Press, 1988). Rex A. Wade includes a comparative geo-

graphic focus in *Red Guards and Workers' Militias in the Russian Revolution* (Stanford, Calif.: Stanford University Press, 1984), but the core of his research remains Petrograd. Ronald Grigor Suny, *The Baku Commune, 1917–1918: Class and Nationality in the Russian Revolution* (Princeton, N.J.: Princeton University Press, 1972), and Donald J. Raleigh, *Revolution on the Volga: 1917 in Saratov* (Ithaca, N.Y.: Cornell University Press, 1986), transcend the Petrograd–Moscow focus. On the preoccupation with Petrograd and Moscow, see Ronald Grigor Suny, "Russian Labor and Its Historians in the West: A Report and Discussion of the Berkeley Conference on the Social History of Russian Labor," *International Labor and Working Class History* 22 (Fall 1982): 39–53, especially 47–51.

4. This now-familiar argument can be found in full in Ronald Grigor Suny, "Toward A Social History of the October Revolution," *American Historical Review* 88 (February 1983), 31–52. The essays in Daniel H. Kaiser, ed., *The Workers' Revolution in Russia, 1917: The View from Below* (Cambridge: Cambridge University Press, 1987), present a well-integrated case for the importance of social history in understanding the 1917 revolution.

5. Diane Koenker, "Moscow in 1917: The View from Below," in *Workers' Revolution in Russia, 1917*, ed. Kaiser, 95, 97; William G. Rosenberg, "Russian Labor and Bolshevik Power: Social Dimensions of Protest in Petrograd after October," in ibid., 128, 131; Suny, "Toward a Social History of the October Revolution," 52.

6. A notable exception, although focused principally on the 1920s, is William J. Chase, *Workers, Society, and the Soviet State: Labor and Life in Moscow, 1918–1929* (Urbana and Chicago: University of Illinois Press, 1987).

7. Robert Service, *The Bolshevik Party in Revolution: A Study in Organisational Change, 1917–1923* (London: Macmillan, 1979), 200–12 and passim; Thomas F. Remington, *Building Socialism in Bolshevik Russia: Ideology and Industrial Organization, 1917–1921* (Pittsburgh: Pittsburgh University Press, 1984), 163, 176–77; Silvana Malle, *The Economic Organization of War Communism, 1918–1921* (Cambridge: Cambridge University Press, 1985), passim, and my review in *Slavic Review* 46 (Spring 1987): 158; Richard Sakwa, *Soviet Communists in Power: A Study of Moscow during the Civil War, 1918–1921* (New York: St. Martin's, 1988), 265–79. T. H. Rigby, *Lenin's Government: Sovnarkom, 1917–1922* (Cambridge: Cambridge University Press, 1979), by definition, intentionally concerns itself with a central focus throughout.

8. M. S. Bernshtam, ed., *Nezavisimoe rabochee dvizhenie v 1918 godu: Dokumenty i materialy* (Paris: YMCA Press, 1981); Vladimir Brovkin, "The Mensheviks' Political Comeback: The Elections to the Provincial City

Soviets in Spring 1918," *Russian Review* 42 (January 1983): 1–50; idem, "Politics, Not Economics Was the Key," *Slavic Review* 44 (Summer 1985): 244–50.

9. K. I. Bobkov, "Iz istorii organizatsii upravleniia promyshlennost'iu v pervye gody sovetskoi vlasti (1917–1920 gg.) (Na materialakh tekstil'noi promyshlennosti)," *Voprosy istorii* 4 (April 1957): 119.

10. G. A. Trukan, *Oktiabr' v tsentral'noi Rossii* (Moscow: Mysl', 1967), 14–15. The region encompassed only 3 percent of the territory of the empire, but its 28.2 million inhabitants, over 97 percent of whom were ethnic Russians, accounted for 21 percent of the total population.

11. *Rabochii klass sovetskoi Rossii v pervyi god diktatury proletariata: Sbornik dokumentov i materialov* (Moscow: Nauka, 1964), 151. L. S. Gaponenko, *Rabochii klass Rossii v 1917 godu* (Moscow: Nauka, 1970), 33–87, lists the total number of workers in Russia as 15 million, but his calculations include hired labor in transport, agriculture, construction, and other nonfactory employment.

12. Jo Ann Ruckman, *The Moscow Business Elite: A Social and Cultural Portrait of Two Generations, 1840–1905* (DeKalb, Ill.: Northern Illinois University Press, 1984), 56–58.

13. L. E. Ankudinova, *Natsionalizatsiia promyshlennosti v SSSR (1917–1920 gg.)* (Leningrad: Izdatel'stvo Leningradskogo Universiteta, 1963), 12.

14. This distinction has been too thoroughly addressed in recent historiography to require recapitulation here. An excellent and thorough treatment is found in McDaniel, *Autocracy, Capitalism, and Revolution in Russia*, 164–212, especially 180, 194. See also Reginald Zelnik's introductory remarks, *A Radical Worker in Tsarist Russia: The Autobiography of Semën Ivanovich Kanatchikov*, trans. and ed. Reginald E. Zelnik (Stanford, Calif.: Stanford University Press, 1986), xv–xxx. The present study also accepts the conclusion, extensively developed in the works by Koenker, Mandel, and Smith already cited, that there existed a direct link between level of skill and degree of political consciousness.

15. Unless otherwise noted, "unskilled" describes those who worked at tasks requiring no particular training, such as wool cleaning and cotton preparation. "Semiskilled" in this instance includes those who had learned functions applicable to textile production but not readily transferable to other spheres, such as the operation of mechanized spindles and looms. "Skilled" will designate those—mechanics, steamfitters—who could readily convert their expertise into employment in other industries.

16. Victoria E. Bonnell, *Roots of Rebellion: Workers' Politics and Organizations in St. Petersburg and Moscow, 1900–1914* (Berkeley and Los Angeles: University of California Press, 1983), 22–23, 32–33.

17. Trukan, *Oktiabr' v tsentral'noi Rossii*, 7.

18. For example, Smith, "Craft Consciousness, Class Consciousness," 43.

19. McDaniel, *Autocracy, Capitalism, and Revolution in Russia*, 302.

20. She notes that within the city of Moscow radical resolutions emerged from textile workers of the Zamoskvorech'e district in 1917, where they worked in close proximity to metalworkers who had passed assertive resolutions earlier in the year. She found no evidence of any resolution favoring Soviet power from the Lefortovo district in northeast Moscow, without radical metalworkers. Koenker, "Moscow in 1917: The View from Below," 90. See also Mandel, *The Petrograd Workers and the Fall of the Old Regime*, 29.

21. *Tekstil'shchik* 17 (June 1920): 3.

22. Vladimir Z. Drobizhev, *Glavnyi shtab sotsialisticheskoi promysh-lennosti: Ocherki istorii VSNKh, 1917–1932* (Moscow: Mysl', 1966), 24. This 85 percent pertains only to textile enterprises within the present borders of the USSR. It excludes Poland, which received political inde-pendence with the collapse of tsarism. Poland had accounted for less than one-fifth of the textile industry of the Russian empire and, by definition, is outside the scope of a study of the organization of Soviet industry.

23. *Piataia Vserossiiskaia konferentsiia professional'nykh soiuzov (3–7 noiabria 1920 g.): Stenograficheskii otchet* (Moscow: [n.p.], 1921), 62.

24. Efim G. Gimpel'son, *Rabochii klass v upravlenii sovetskim gosu-darstvom, noiabr' 1917–1920 gg.* (Moscow: Nauka, 1982), 273. Such state-ments require qualification, since these were gains only relative to other industries. The massive exodus of managerial and technical personnel, for example, goes further toward explaining the statistical progress toward workers' management than any meaningful acquisition of the requisite expertise by workers and their representatives.

25. *Sbornik dekretov i postanovlenii po narodnomu khoziaistvu*, vol. 3 (Moscow: [n.p.], 1921), 389–90; *Tekstil'shchik* 9–10 (25 December 1918): 4. By 1920–1921, the Red Army received 40 percent of the output of cotton goods and 70–100 percent of other textile products. Iu. Poliakov, *The Civil War in Russia* (Moscow: Progress Publishers, 1981), 92.

26. Lenin, *PSS*, vol. 36, 465–66; *Plenum Vysshego Soveta Narodnogo Khoziaistva, 14–23 sentiabria 1918 goda (Stenograficheskii otchet)* (Mos-cow: VSNKh, 1918), 8, 118, 130; L. V. Strakhov, "Natsionalizatsiia krup-noi promyshlennosti goroda Moskvy," *Uchenye zapiski Moskovskogo in-stituta im. V. I. Lenina*, no. 200 (1964): 281.

27. On the differences between available published sources for central and outlying areas, refer to Rosenberg and Koenker, "Skilled Workers and the Strike Movement in Revolutionary Russia," 624.

28. For an important exchange on this set of issues, see Sheila Fitzpatrick, "The Bolsheviks' Dilemma: Class, Culture, and Politics in the Early Soviet Years," *Slavic Review* 47 (Winter 1988), 599–613; Ronald Grigor Suny, "Class and State in the Early Soviet Period: A Reply to Sheila Fitzpatrick," ibid., 614–19; Daniel Orlovsky, "Social Histroy and Its Categories," ibid., 620–23; Sheila Fitzpatrick, "Reply to Suny and Orlovsky," ibid., 624–26.

29. J. Arch Getty, *Origins of the Great Purges: The Soviet Communist Party Reconsidered, 1933–1938* (Cambridge: Cambridge University Press, 1985); Lynne Viola, *The Best Sons of the Fatherland: Workers in the Vanguard of Soviet Collectivization* (New York: Oxford University Press, 1987); Alan M. Ball, *Russia's Last Capitalists: The Nepmen, 1921–1929* (Berkeley and Los Angeles: University of California Press, 1987); Hiroaki Kuromiya, *Stalin's Industrial Revolution: Politics and Workers, 1928–1932* (Cambridge: Cambridge University Press, 1988); Anne Rassweiler, *The Generation of Power* (New York: Oxford University Press, 1988); Lewis H. Siegelbaum, *Stakhanovism and the Politics of Productivity in the USSR, 1935–1941* (Cambridge: Cambridge University Press, 1988). See also Moshe Lewin, *The Making of the Soviet System: Essays in the Social History of Interwar Russia* (New York: Pantheon Books, 1985); Sheila Fitzpatrick, *Education and Social Mobility in the Soviet Union, 1921–1934* (Cambridge: Cambridge University Press, 1978); idem, *The Russian Revolution* (Oxford and New York: Oxford University Press, 1982); and idem, ed., *Cultural Revolution in Russia, 1928–1931* (Bloomington, Ind.: Indiana University Press, 1978).

Chapter 2

1. N. P. Langovoi, "Manufakturnaia promyshlennost'," in *Fabrichno-zavodsakaia promyshlennost' i torgovli Rossii* (St. Petersburg: Izdanie Departamenta Torgovli i Manufaktur Ministerstva Finansov, 1893), 1.

2. E. I. Zaozerskaia, "Manufaktura v seredine XVIII veka," *Istoricheskie zapiski* 33 (1950): 125, 128.

3. A. G. Rashin, *Formirovanie rabochego klassa Rossii: Istoriko-ekonomicheskie ocherki* (Moscow: Izdatel'stvo sotsial'no-ekonomicheskoi literatury, 1958), 12–16 ff.

4. A. M. Korneev, *Tekstil'naia promyshlennost' SSSR i puti ee razvitiia* (Moscow: Izdatel'stvo literatury po legkoi promyshlennosti, 1957), 45.

5. Mikhail I. Tugan-Baranovsky, *The Russian Factory in the 19th Century*, trans. Arthur Levin and Claora Levin (Homewood, Ill.: Richard D. Irwin, Inc., 1970), 48–49, 55–57, 301, 365–71; K. A. Pazhitnov, *Ocherki*

istorii tekstil'noi promyshlennosti dorevoliutsionnoi Rossii: Sherstianaia promyshlennost' (Moscow: Izdatel'stvo Akademii nauk SSSR, 1958), 82, 126, 130–47; Robert E. Johnson, *Peasant and Proletarian: The Working Class of Moscow in the Late Nineteenth Century* (New Brunswick, N.J.: Rutgers University Press, 1979), 12–20.

6. Johnson, *Peasant and Proletarian*, 35.

7. As late as 1911, 83 percent of the 61,000 textile workers of Kostroma province worked outside urban areas (Olga Crisp, *Studies in the Russian Economy before 1914* [London: Macmillan, 1976], 45). In 1912, only 31.2 percent of Russia's wool factories were located in towns or cities. (Pazhnitnov, *Ocherki tekstil'noi promyshlennosti*, 188).

8. V. M. Selunskaia, *Izmeneniia sotsial'noi struktury sovetskogo obshchestva, oktiabr' 1917–1920* (Moscow: Mysl', 1976), 137–38. By 1917, 40.1 percent of the textile workers of Vladimir province and 37.4 percent of those in Moscow province maintained rural ties. G. A. Trukan, "O nekotorykh voprosakh rabochego dvizheniia v tsentral'nom promyshlennom raione (fevral'–oktiabr' 1917 g.)," in *Rabochii klass i rabochee dvizhenie v Rossii v 1917 g.* (Moscow: Nauka, 1964), 106.

9. Pazhitnov, *Ocherki tekstil'noi promyshlennosti*, 56–57.

10. Rose Glickman, *Russian Factory Women: Society and Workplace, 1880–1914* (Berkeley: University of California Press, 1984), passim.

11. *Tekstil'shchik* 17 (June 1920): 6.

12. While these complaints are virtually universal in worker memoirs, see especially V. P. Nogin, *Fabrika Palia* (Leningrad: Pvorka, 1924), 20–24.

13. Pazhitnov, *Ocherki tekstil'noi promyshlennosti*, 170, 210–11; R. P. Mikhail'kov, *Ocherki istorii Trekhgornoi manufaktury v sviazi s istoriei tekstil'noi promyshlennosti za 170 let* (Rostov-on-Don: [n.p.], 1972), 202–03.

14. Mikhail'kov, *Ocherki istorii Trekhgornoi manufaktury*, 199–203.

15. In 1908, 25.3 percent of the women and 72.5 percent of the men in the cotton industry were literate. By 1918, the literacy rate for women throughout the industry rose to 37.5 percent. See Rose Glickman, "The Russian Factory Woman, 1880–1914," in Dorothy Atkinson, Alexander Dallin, and Gail Warshovsky Lapidus, eds., *Women in Russia* (Stanford, Calif.: Stanford University Press, 1977), 73. This increase was due to a literacy rate two to three times greater among young girls than older women. See *Teksti'shchik* 9–10 (25 December 1918): 9. Personal profiles of women who entered the work force in the twentieth century contain a few years of schooling, in contrast to their older counterparts. See Ol'ga Nestorova Chaadaeva, ed., *Rabotnitsa na sotsialisticheskoi stroike: Sbornik avtobiografii rabotnits* (Moscow: Partiinoe izdatel'stvo, 1932).

16. S. Lapitskaia, *Byt rabochikh Trekhgornoi manufaktury* (Moscow: Istoriia zavodov, 1935), 74.

17. For example, Pazhitnov, *Ocherki tekstil'noi promyshlennosti*, 175–80; Allan K. Wildman, *The Making of a Workers' Revolution* (Chicago: University of Chicago Press, 1967), 23–27, 55–56, 65, 246; McDaniel, *Autocracy, Capitalism, and Revolution in Russia*, 155, 178, 238.

18. Pazhitnov, *Ocherki tekstil'noi promyshlennosti*, 217.

19. F. A. Romanov, *Tekstil'shchiki Moskovskoi oblasti v gody grazhdanskoi voiny* (Moscow: Profizdat, 1939), 6; McDaniel, *Autocracy, Capitalism, and Revolution in Russia*, 302.

20. Johnson, *Peasant and Proletarian*, 117–18, 146–47.

21. *Tekstil'nyi rabochii* 1 (5 September 1917): 7–8; *Professional'nye soiuzy rabochikh Rossii, 1905g.–fevral' 1917 g.: Perechen' organizatsii*, vol. 1 (Moscow: Nauka, 1985): 64–73, 138–43, 170–98.

22. McDaniel, *Autocracy, Capitalism, and Revolution in Russia*, 291–92; M. A. Abashkina et al., *Povest' o trekh* (Moscow: Profizdat, 1935), 16, 33.

23. "Valentina Ivanovna Petrova," in *Rabotnitsa na sotsialisticheskoi stroike*, 45–46; Abashkina, *Povest' o trekh*, 43.

24. G. Tsiperovich, *Sindikaty i tresty v dorevoliutsionnoi Rossii i v SSSR, izdanie chertvertoe* (Leningrad: Izdatel'stvo tekhnika i proizvodsto, 1927); P. V. Volobuev, *Proletariat i burzhuazii v Rossii v 1917 g.* (Moscow: Mysl', 1964); Ruckman, *The Moscow Business Elite*; Lewis H. Siegelbaum, *The Politics of Industrial Mobilization in Russia, 1914–1917: A Study of the War-Industries Committees* (New York: St. Martin's, 1983); Alfred J. Rieber, *Merchants and Entrepreneurs in Imperial Russia* (Chapel Hill, N.C.: University of North Carolina Press, 1982); Muriel Joffe, "The Cotton Manufacturers in the Central Industrial Region, 1880s–1914: Merchants, Economics, and Politics," Ph.D. diss., University of Pennsylvania, 1981; Thomas C. Owen, *Capitalism and Politics in Russia: A Social History of the Moscow Merchants, 1855–1905* (Cambridge: Cambridge University Press, 1981).

25. Tsiperovich, *Sindikaty i tresty*, 247–49; Joffe, *The Cotton Manufacturers in the Central Industrial Region*, 280–300; V. Ia. Laverychev, "Sozdanie tsentral'nykh gosudarstvennykh organov upravleniia tekstil'noi promyshlennosti v 1918 g. (iz istorii 'Tsentrotekstilia')," in *Iz istorii Velikoi Oktiabr'skoi revoliutsii* (Moscow: Izdatel'stvo Moskovskogo universiteta, 1957), 114; *Glavtekstil': Kratkii otchet glavnogo pravleniia tekstil'nykh predpriiatii RSFSR* (Moscow: [n.p.], 1920), 5; G. S. Ignat'ev, *Moskva v pervyi god proletarskoi diktatury* (Moscow: Nauka, 1975), 146. Despite the general Soviet characterization of the textile industry as highly monopolized, only 40 percent of its enterprises were part of larger combines by 1913. This was the lowest rate for any major industry in the country and the only Russian industry less concentrated than its German counterpart (Ankudinova, *Natsionalizatsiia promyshlennost' v SSSR*, 12).

26. Crisp, *Studies in the Russian Economy before 1914*, 142–43; Ruckman, *The Moscow Business Elite*, 56–58.

27. The comparison is in terms of 1913 rubles. V. P. Miliutin, *Istoriia
ekonomicheskogo razvitiia SSSR, 1917–1927* (Moscow: Gosizdat, 1928),
56–61; S. O. Zagorsky, *State Control of Industry in Russia during the War*
(New Haven: Yale University Press, 1928), 27–29.

28. Zagorsky, *State Control of Industry in Russia*, 131–56; Laverychev,
"Sozdanie organov upravleniia tekstil'noi promyshlennosti," 114–16;
Glavtekstil', 7–9, 17–18; Tsiperovich, *Sindikaty i tresty*, 310; Siegelbaum,
The Politics of Industrial Mobilization in Russia, 142–47; V. Z. Drobizhev,
"Obrazovanie sovetov narodnogo khoziaistva v Moskovskom promyshlen-
nom raione (1917–1918 gg.)," in *Iz istorii Velikoi Oktiabr'skoi revoliutsii*,
84. By 1916, however, problems in attaining raw materials and in transport
led to a serious decline in output. See Zagorsky, *State Control of Industry
in Russia*, 36–44, 245–49; Korneev, *Tekstil'naia promyshlennost' SSSR*, 55;
Romanov, *Tekstil'shchiki Moskovskoi oblasti*, 48.

29. *Fabrichno-zavodskaia promyshlennost' v period 1913–1918 gg.:
Vserossiiskaia promyshlennaia i professional'naia perepis' 1918 g., vypusk
I* (Moscow: Tsentral'noe statisticheskoe upravlenie, 1926), 100.

30. Trukan, *Oktiabr' v tsentral'noi Rossii*, 82.

31. E. N. Burdzhalov, *Russia's Second Revolution: The February 1917
Uprising in Petrograd*, trans. and ed. Donald J. Raleigh (Bloomington and
Indianapolis: Indiana University Press, 1987), 23; G. K. Korolev, *Ivanovo–
Kineshemskie tekstil'shchiki v 1917 godu (iz vospominanii tekstil'shchika)*
(Moscow: Izdatel'stvo VTsSPS, 1927), 13.

32. See, for example, "Valentina Ivanovna Petrova," in *Rabotnitsa na
sotsialisticheskoi stroike*, 45; "Anna Aleksandrovna Guliutina," in ibid.,
69.

33. Koenker, *Moscow Workers and the 1917 Revolution*, 198, 211, 257–
58.

34. This section confines description to that necessary for the analysis of
the contours of the workers' movement that follows. Fuller detail is available in English and Russian from the works consulted in compiling this
section. Those in English were cited in the notes to Chapter 1. The most
relevant Soviet works include E. N. Burdzhalov, *Vtoraia russkaia revoliutsiia*, vol. 2. *Moskva. Front. Periferiia.* (Moscow: Nauka, 1971); Trukan,
Oktiabr' v tsentral'noi Rossii; A. Ia. Grunt, *Moskva 1917-i: Revoliutsiia i
kontrrevoliutsiia* (Moscow: Nauka, 1976); Volobuev, *Proletariat i burzhuaziia*; L. S. Gaponenko, *Rabochii klass Rossii v 1917 godu*; P. A. Nikolaev,
*Rabochie-metallisty tsentral'no-promyshlennogo raiona Rossii v bor'be
za pobedu Oktiabr'skoi revoliutsii (mart–noiabr' 1917 g.)* (Moscow:
Izdatel'stvo VPSH i AON pri TsK KPSS, 1960); *Istoriia rabochikh*

Moskvy, 1917–1925 gg. (Moscow: Nauka, 1983); *Rabochii klass i rabochee dvizhenie v Rossii v 1917 g.; Rabochii klass v Oktiabr'skoi revoliutsii i na zashchite ee zavoevanie, 1917–1920 gg.*, vol. 1 (Moscow: Nauka, 1984).

35. In 1917, the administrative borders of Vladimir and Kostroma provinces were anachronistic, and the industrial contours of this key textile-producing region had long determined its economic and political identity. The Ivanovo–Kineshma Region in 1917 consisted of all of the Kineshma and Iur'evets districts and sixteen *volosty* of the Nerekhta district of Kostroma province; the Shuia district and eight *volosty* each from the Suzdal' and Kovrov districts of Vladimir province. In January 1918, the Second Congress of Soviets of the Ivanovo–Kineshma Region formed a commission under M. V. Frunze to work for administrative reorganization, and on June 20 the People's Commissariat of Internal Affairs designated the region Ivanovo–Voznesensk province. The area is sometimes referred to as the Ivanovo–Voznesensk Region after its principal city (now Ivanovo). Herein Ivanovo–Kineshma will denote the region and Ivanovo–Voznesensk the city in the period prior to June 20, 1918. For the period following the creation of Ivanovo–Voznesensk province, the text will distinguish between the city and province.

36. For example, in Teikovo village (Vladimir province), local authorities withheld the news of the tsar's fall until March 3 by controlling the post and telegraph and circulating no newspapers. V. A. Babichev, I. I. Zimin, and V. M. Smirnov, *Teikovskii khlopchatobumazhnyi: Istoricheskii ocherk* (Iaroslavl: Verkhne-Volzhskoe knizhnoe izdatel'stvo, 1966), 28.

37. *Sed'maia (aprel'skaia) Vserossiiskaia konferentsiia (bol'shevikov): Protokoly* (Moscow: Gosudarstvennoe izdatel'stvo politicheskoi literatury, 1958), 134. Quoted in McDaniel, *Autocracy, Capitalism, and Revolution in Russia*, 349.

38. *Shestoi s"ezd RSDRP (bol'shevikov): Avgust 1917 goda. Protokoly* (Moscow: Gosudarstvennoe izdatel'stvo politicheskoi literatury, 1958), 334.

39. Ibid., 331.

40. Korolev, *Ivanovo–Kineshemskie tekstil'shchiki*, 59.

41. Nikolaev, *Rabochie-metallisty tsentral'no-promyshlennogo raiona*, 18–103.

42. Diane Koenker, "The Evolution of Party Consciousness in 1917: The Case of the Moscow Workers," *Soviet Studies* 30 (1978): 38–62.

43. *KPSS v rezoliutsiiakh i resheniiakh s"ezdov, konferentsii i plenumov TsK*, vol. 1 (Moscow: Gosudarstvennoe izdatel'stvo politicheskoi literatury, 1970), 61–65.

44. Lenin, *PSS*, vol. 31, 58–59.

45. Ibid., 116. Lenin's italics.

46. Ibid., 168, 412, 414.

47. Workers' control (*kontrol'*) in this sense entailed establishing accounting and supervisory authority over production until the workers acquired the expertise to implement full workers' management.

48. Paul Avrich, "The Russian Revolution and the Factory Committees" (Ph.D. diss., Columbia University, 1961), 162–66. Lenin supported the third of these alternatives.

49. *Protokoly shestogo s"ezda RSDRP(b)* (Moscow: Partiinoe izdatel'stvo, 1934), 242–43.

50. Lenin, *PSS*, vol. 34, 309.

51. Ibid., 305–06, 311–12, 320; vol. 33, 49.

52. For an extended study of the centralizing proclivities included in various Bolshevik positions, see Malle, *The Economic Organization of War Communism*. A good sense of the political divisions at the top of the party can be found in D. A. Longley, "The Divisions in the Bolshevik Party in March 1917," *Soviet Studies* 24 (July 1972): 61–76.

53. A. I. Mel'chin, "Nekotorye voprosy partiinogo stroitel'stva posle Oktiabria," in *Iz istorii grazhdanskoi voiny i interventsii, 1917–1922 gg.* (Moscow: Nauka, 1974), 66.

54. Service, *The Bolshevik Party in Revolution*, 85–111.

55. Quoted in Leonard Schapiro, *The Communist Party of the Soviet Union* (New York: Vintage Books, 1971), 247; see also T. H. Rigby, *Communist Party Membership in the U.S.S.R.* (Princeton, N.J.: Princeton University Press, 1968), 74–75.

56. Schapiro, *The Communist Party of the Soviet Union*, 249.

57. A. A. Timofeevski et al., *V. I. Lenin i stroitel'stvo partii v pervye gody sovetskoi vlasti* (Moscow: Mysl', 1965), 72.

58. V. Ia. Laverychev and A. M. Soloveva, *Boevoi pochin rossiiskogo proletariata: K 100-letiiu Morozovskoi stachki 1885 g.* (Moscow: Mysl', 1985); V. Sokolov, "Stachka tkachei Ivanovo–Voznesenskoi manufaktury v 1895 g.," *Krasnyi arkhiv*, vol. 5, no. 72 (1935): 178–83; William G. Gard, "The Party and the Proletariat in Ivanovo–Voznesensk, 1905," *Russian History/Histoire Russe*, vol. 2, no. 2 (1975): 101–23; P. M. Ekzempliarskii, *Istoriia goroda Ivanova, chast' I* (Ivanovo: Ivanskoe knizhnoe izdatel'stvo, 1958); *Pervyi sovet rabochikh deputatov: Vremia' sobitiia, liudi. Ivanovo–Vozneseksk, 1905* (Moscow: Sovetskaia Rossiia, 1985); F. N. Samoilov, "Oktiabr'skaia revoliutsiia v Ivanovo–Voznesenske," in *Pobeda Velikoi Oktiabr'skoi sotsialisticheskoi revoliutsii: Sbornik vospominanii* (Moscow: Izdatel'stvo politicheskoi literatury, 1958), 1–7.

59. Korolev, *Ivanovo–Kineshemskie tekstil'shchiki*, 13, 20, 29.

60. Owen, *Capitalism and Politics in Russia*; Rieber, *Merchants and*

Entrepreneurs in Imperial Russia; Siegelbaum, *The Politics of Industrial Mobilization in Russia*; Ruckman, *The Moscow Business Elite*; James L. West, "The Riabushinskij Circle: Russian Industrialists in Search of a Bourgeoisie, 1909–1914," *Jahrbücher für Geschichte Osteuropas*, Band 32, Heft 3 (1984): 358–77; Joffe, *The Cotton Manufacturers of the Central Industrial Region*; Volobuev, *Proletariat i burzhuaziia*, 50.

61. Volobuev, *Proletariat i burzhuaziia*, 49–55, 74–75. Five *raion* branches existed in February 1917. By September, the number grew to twenty, and all major industrial regions were represented. The All-Russian Union of Factory and Works Owners officially formed on August 9.

62. Ibid., 71.

63. Ibid., 74; *Materialy po istorii SSSR*, vol. 3 (Moscow: Izdatel'stvo Akademii nauk SSSR, 1956), 48, n. 1. In September 1917, this organization merged with the Society of Factory Owners of the Cotton Industry. By January 1918, the union encompassed fourteen *raion* societies whose members owned 164 enterprises employing 311,000 workers. The Central Industrial Region consisted of Moscow, Iaroslav, Kaluga, Nizhnii-Novgorod, Riazan, Tula, and Tver provinces in addition to Vladimir and Kostroma.

64. Korolev, *Ivanovo-Kineshemskie tekstil'shchiki*, 25–27.

65. *Tekstil'nyi rabochii* 5 (20 December 1917): 7; Laverychev, "Sozdanie organov upravleniia tekstil'noi promyshlennosti," 116.

66. Robert Paul Browder and Alexander Kerensky, eds., *The Provisional Government: Documents*, vol. 2 (Stanford, Calif.: Stanford University Press, 1961), 718–20.

67. V. Petrova, "Likinskie tekstil'shchiki," in *Za vlast' sovetov* (Moscow: Moskovskii rabochii, 1957), 418–19.

68. *Za vlast' sovetov: Sbornik dokumentov i vospominanii* (Ivanovo: Verkhne-Volzhskoe knizhnoe izdatel'stvo), 88–89; V. M. Sokolov, *Fabrika imeni S. I. Balashova* (Ivanovo: Ivanskoe knizhnoe izdatel'stvo, 1961), 33–34; *Materialy po istorii SSSR*, 31–32. This mill was also nationalized in January 1918.

69. *Za vlast' sovetov. Sbornik dokumentov i vospominanii*, 156–57.

70. For example, *Materialy po istorii SSSR*, 42–44, 46, 51–52, 59–60, 61.

71. *Tekstil'nyi rabochii* 1 (5 September 1917): 9–11.

72. Strakhov, "Natsionalizatsiia krupnoi promyshlennosti goroda Moskvy," 267.

73. Romanov, *Tekstil'shchiki Moskovskoi oblasti*, 48.

74. *TsGAOR* (Central State Archive of the October Revolution), f. 5457, op. 1, d. 54, l. 3.

75. *TsGAOR*, f. 5457, op. 1, d. 54, l. 1.

76. *Materialy po istorii SSSR*, 43–44.

77. S. M. Klimokhin, *Kratkaia istoriia stachki tekstil'shchikov Ivanovo–Kineshemskoi promyshlennoi oblasti* (*s 21-go oktiabria po 17-e noiabria 1917 g.*) (Kineshma: Tipografiia khoziastv. sektsiia SRSiKD, 1918), 1.

78. Trukan, *Oktiabr' v tsentral'noi Rossii*, 82–87.

79. *Za vlast' sovetov: Sbornik dokumentov i vospominanii*, 157; Trukan, *Oktiabr' v tsentral'noi Rossii*, 87; *Pravda*, 29 October 1917, 3; *Rabochii klass v Oktiabr'skoi revoliutsii*, 87; Ekzempliarskii, *Istoriia goroda Ivanova, chast' I*, 348.

80. *Tekstil'shchik* 17 (June 1920), 25; M. Frunze, "Ob obrazovanii novoi Ivanovo–Voznesenskoi gubernii iz chastei Kostromsk. i Vladimirsk.," in *Sbornik statei i materialov po obrazovanii Ivanovo–Voznesenskoi gubernii. Vypusk 1* (Ivanovo–Voznesensk: Izdanie I-V gubispolkom, 1918), 5. The average textile factory in the Central Industrial Region employed 510 workers in 1917. Even taking intervening closings into account, the industrial census of August 31, 1918, counted three operating textile factories in Vladimir province whose work force exceeded 5,000 per enterprise. An additional eight factories employed 2,501–5,000, sixteen employed 1,001–2,500, twenty employed 501–1,000, and thirty-two counted 201–500. In Kostroma province, the census reported one factory that employed more than 5,000, three in the 1,001–2,500 category, and four employing 210–500. *Fabrichno-zavodskaia promyshlennost' v period 1913–1918 gg.: Vserossiiskaia promyshlennaia i professional'naia perepis' 1918 g., vypusk 2* (Moscow: Tsentral'noe statisticheskoe upravlenie, 1926), 30–31.

81. Examples would include Shuia and Kineshma as well as the industrial villages of Vichuga, Rodniki, Sereda, Pistsovo, Dulianino, Kovrov, Kameshki, Tyntsovo, and Kohkma. Vichuga, for example, employed 35,000 workers in eight textile enterprises. Burdzhalov, *Vtoraia russkaia revoliutsiia*, vol. 2. *Moskva. Front. Periferiia.*, 184. The exception was the town of Vladimir, about one-fifth the size of Ivanovo–Voznesensk, which did not possess a significant concentration of large-scale textile mills.

82. Early in 1917, a Menshevik and a Socialist-Revolutionary addressed a general meeting in a Bogorodsk mill. Uncertain of their response, the workers decided that those who still held land should support the SRs and those without would back the Mensheviks. "Klavdia Sevast'iannovna Ukolova," in *Rabotnitsa na sotsialisticheskoi stroike*, 62.

83. Korolev, *Ivanovo–Kineshemskie tekstil'shchiki*, 32–34.

84. Klimokhin, *Kratkaia istoriia stachki tekstil'shchikov*, 1–2; *Pravda*, 29 October 1917, 3.

85. *Pravda*, 29 October 1917, 3.

86. Trukan, "O nekotorykh voprosakh rabochego dvizheniia," 113.

87. V. M. Kurakhtanov, *Pervaia sittsenabivaniia* (Moscow: Sotsial'no-ekonomicheskoi literatury, 1960), 56.

88. Gaponenko, *Rabochii klass Rossii v 1917 godu*, 286.

89. Trukan, "O nekotorykh voprosakh rabochego dvizheniia," 113–14.

90. *Tekstil'nyi rabochii* 3 (26 October 1917): 2–10. Of the eighty-three representatives, fifty were Bolsheviks. Under tsarist rule, thirty-eight had served prison sentences for revolutionary activity and twenty-five had been under police surveillance. Trukan, "O nekotorykh voprosakh rabochego dvizheniia," 114.

91. *Za vlast'sovetov: Sbornik dokumentov i vospominanii*, 158–60.

92. *Moskva, Oktiabr', Revoliutsiia: Dokumenty i vospominaniia* (Moscow: Moskovskii rabochii, 1957), 156; Trukan, *Oktiabr' v tsentral'noi Rossii*, 215; Volobuev, *Proletariat i burzhuaziia*, 240. These delegates represented more than 200,000 textile workers.

93. Korolev, *Ivanovo–Kineshemskie tekstil'shchiki*, 37–38; Klimokhin, *Kratkaia istoriia stachki tekstil'shchikov*, 3–4. The Bolsheviks, who already held five of the seven offices on the union board, assumed seven of the nine posts on the Central Strike Committee.

94. *Revoliutsionnoe dvizhenie v Rossii nakanune Oktiabr'skogo vooruzhennogo vosstanie (1–24 oktiabria 1917 g.)* (Moscow: Izdatel'stvo Akademii nauk SSSR, 1962), 291–92.

95. *Za vlast'sovetov: Sbornik dokumentov i vospominanii*, 165–69.

96. Rosenberg and Koenker, "Skilled Workers and the Strike Movement in Revolutionary Russia," 614–15.

97. *Revoliutsionnoe dvizhenie v Rossii*, 297–99.

98. *Moskva, Oktiabr', Revoliutsiia*, 168.

99. A. M. Lisetskii, *Bol'sheviki vo glave massovykh stachek* (Kishinev: Izdatel'stvo Shtiintsa, 1974), 129–30.

100. *Profsoiuzy SSSR: Dokumenty i materialy, vol. I (1905–1917 gg.)* (Moscow: Profizdat, 1963), 464–68; *Za vlast'sovetov: Sbornik dokumentov i vospominanii*, 174–75.

101. Korolev, *Ivanovo–Kineshemskie tekstil'shchiki*, 49–56.

102. *Revoliutsionnoe dvizhenie v Rossii*, 323.

103. *Revoliutsionnoe dvizhenie v Rossii*, 327; *Materialy po istorii SSSR*, 70–71.

104. *Materialy po istorii SSSR*, 48.

105. *Za vlast'sovetov: Sbornik dokumentov i vospominanii*, 180.

106. Korolev, *Ivanovo–Kineshemskie tekstil'shchiki*, 60.

107. *Sotsial-Demokrat*, 29 October 1917, 2.

108. *Sotsial-Demokrat*, 3 November 1917, 1.

109. *Moskva, Oktiabr', Revoliutsiia*, 247–48.

110. *Tekstil'nyi rabochii* 4 (9 December 1917): 9.
111. Korolev, *Ivanovo–Kineshemskie tekstil'shchiki*, 62–63.
112. Klimokhin, *Kratkaia istoriia stachki tekstil'shchikov*, 40–41.
113. *Sotsial-Demokrat*, 23 November 1917, 3.
114. Korolev, *Ivanovo–Kineshemskie tekstil'shchiki*, 64–65.
115. *Za vlast'sovetov: Sbornik dokumentov i vospominanii*, 310, 313–14.
116. *Tekstil'nyi rabochii* 4 (9 December 1917): 4; *Tekstil'nyi rabochii* 5 (20 December 1917): 2–3; *Ivanovo–Voznesenskie bol'sheviki v period podgotovki i provedeniia Velikoi Oktiabr'skoi revoliutsii: Sbornik dokumentov* (Ivanovo: Ivanskoe oblastnoe gosudarstvennoe izdatel'stvo, 1947), 178; *Rabochii klass v Oktiabr'skoi revoliutsii*, 192–93; *Materialy po istorii SSSR*, 88–89; Korolev, *Ivanovo–Kineshemskie tekstil'shchiki*, 64–66; S. S. Deev et al., eds., *Istoriia Ivanova, chast' II* (Ivanovo: Ivanskoe knizhnoe izdatal'stvo, 1962), 25–27.
117. Korolev, *Ivanovo–Kineshemskie tekstil'shchiki*, 67–68.
118. An institution that the revolutionary government created in December 1917 to assume eventual complete direction of the national economy.
119. *Tekstil'nyi rabochii* 5 (20 December 1917): 9–10.
120. *Rabochii krai* 95 (16 March 1918): 3.
121. Korolev, *Ivanovo–Kineshemskie tekstil'shchiki*, 68.
122. *Materialy po istorii SSSR*, 81–82, 84–86, 91–93; *Za vlast'sovetov: Sbornik dokumentov i vospominanii*, 314–17.
123. *Sbornik dekretov i postanovlenii po narodnomu khoziaistvu, vypusk 1 (25 oktiabria 1917 g.–25 oktiabria 1918 g.)* (Moscow: [n.p.], 1918), 171–72.
124. *Sbornik uzakonenii i rasporiazhenii rabochago i krest'ianskago pravitel'stva* [hereafter *SU*], 5 (16 December 1917): 73–74.
125. *SU* 3 (8 December 1917): 38; *SU* 9 (24 December 1917): 134–35.
126. *SU* 9 (24 December 1917): 136.
127. *Protokoly I-go Vserossiiskogo s"ezda professionali'nykh soiuzov tekstil'shchikov i fabrichnykh komitetov* (Moscow: Izdanie Vserossiikogo soveta professional'nykh soizov tekstil'shchikov, 1918), 18.
128. Laverychev, "Sozdanie organov upravleniia tekstil'noi promyshlennosti," 120.
129. *Tekstil'nyi rabochii* 5 (20 December 1917): 10–11.
130. Ibid., 15–16.
131. Romanov, *Tekstil'shchiki Moskovskoi oblasti*, 48–50.
132. Laverychev, "Sozdanie organov upravleniia tekstil'noi promyshlennosti," 120.
133. Ibid., 120–21.

134. *Tekstil'nyi rabochii* 5 (20 December 1917): 1.
135. *Natsionalizatsiia promyshlennosti v SSSR: Sbornik dokumentov i materialov, 1917–1920 gg.* (Moscow: Gosudarstvennoe izdatel'stvo politicheskoi literatury, 1954), 93–95.
136. Korolev, *Ivanovo–Kineshemskie tekstil'shchiki*, 69.

Chapter 3

1. Quoted in Evan Mawdsley, *The Russian Civil War* (Boston: Allen & Unwin, 1987), 32.
2. *Resheniia partii i pravitel'stva po khoziaistvennym voprosam, vol. I (1917–1928 gody)* (Moscow: Gosudarstvennoe izdatel'stvo politicheskoi literatury, 1967), 25–27.
3. *Tekstil'nyi rabochii* 5 (20 December 1917): 5.
4. *Narodnoe khoziaistvo* 3 (May 1918): 27.
5. *TsGAOR*, f. 5457, op. 2, d. 5, ll. 9–10.
6. William G. Rosenberg, "Workers and Workers' Control in the Russian Revolution," *History Workshop* 5 (Spring 1978): 91–92. See also Thomas Remington, "Institution Building in Bolshevik Russia: The Case of *Kontrol'*," *Slavic Review* 41 (Spring 1982): 95; Avrich, "The Russian Revolution and the Factory Committees," 49.
7. *Narodnoe khoziaistvo* 5 (5 July 1918): 6. Exact statistics vary but overwhelmingly support the approximations cited here. See Miliutin, *Istoriia ekonomicheskogo razvitiia SSSR, 1917–1927*, 109–10; L. N. Kritsman, *Geroicheskii period velikoi russkoi revoliutsii* (Moscow: Gosudarstvennoe izdatel'stvo, 1924), 41; Drobizhev, *Glavnyi shtab sotsialisticheskoi promyshlennosti*, 99; Alexander Baykov, *The Development of the Soviet Economic System* (Cambridge: Cambridge University Press, 1946), 5 n.; Maurice Dobb, *Soviet Economic Development since 1917* (New York: International Publishers, 1948), 90; J. L. H. Keep, *The Russian Revolution: A Study in Mass Mobilization* (London: Weidenfield and Nicolson, 1976), 266–67.
8. *Natsionalizatsiia promyshlennosti SSSR: Sbornik dokumentov i materialov, 1917–1920 gg.*, 303–04. On November 17, 1917, the Council of People's Commisars made the Likino Mill the first industrial enterprise nationalized by central decree after the revolution. See *SU* 37 (27 May 1918): 458. *Sovnarkom* also soon nationalized the Ivanovo–Voznesensk Weaving Mill on January 27, 1918. Sokolov, *Fabrika imeni S. I. Balashova*, 38.
9. *Natsionalizatsiia promyshlennosti SSSR: Sbornik dokumentov i materialov, 1917–1920 gg.*, 472–73.

10. Ibid., 155–56, 365–70, 374–77.

11. On unregistered nationalizations, see Remington, *Building Social-ism in Bolshevik Russia*, 56–58.

12. A. V. Venediktov, *Organizatsiia gosudarstvennoi promyshlennosti v SSSR, vol. I (1917–1920)* (Leningrad: Izdatel'stvo Leningradskogo univer-siteta, 1957), 180–81; Ankudinova, *Natsionalizatsiia promyshlennost' v SSSR*, 48.

13. Lenin, *PSS*, vol. 36, 175. Lenin's italics.

14. Ibid., 176–77.

15. Ankudinova, *Natsionalizatsiia promyshlennost' v SSSR*, 49–50; V. Z. Drobizhev, "Bor'ba russkoi burzhuazii protiv natsionalizatsii pro-myshlennosti v 1917–1920 gg.," *Istoricheskie zapiski* 68 (1961): 29, 39–41; Kritsman, *Geroicheskii period velikoi russkoi revoliutsii*, 41; Strakhov, "Natsionalizatsiia krupnoi promyshlennosti goroda Moskvy," 222.

16. Ankudinova, *Natsionalizatsiia promyshlennost' v SSSR*, 50; Dro-bizhev, "Bor'ba russkoi burzhuazii protiv natsionalizatsii promyshlen-nosti," 41.

17. E. H. Carr, *The Bolshevik Revolution*, vol. 2 (London: Macmillan, 1952), 87.

18. *Narodnoe khoziaistvo* 2 (April 1918): 1–2.

19. Lenin, *PSS*, vol. 36, 218.

20. Ibid., 295. Lenin's italics.

21. See *Kommunist* 2 (1918): 5; *Kommunist* 3 (1918): 11.

22. N. Osinskii [V. V. Obolenskii], *Stroitel'stva sotsializma* (Moscow: Kommunist, 1918), 25.

23. *SU* 34 (4 May 1918): 9–10.

24. Lenin, *PSS*, vol. 36, 349.

25. *TsGAOR*, f. 5457, op. 2, d. 7, l. 47.

26. *SU* 45 (23 June 1918): 551–52.

27. V. V. Zhuravlev, *Dekrety sovetskoi vlasti kak istoricheskii istochnik* (Moscow: Nauka, 1979), 268–72.

28. *Trudy I Vserossiiskogo s"ezda sovetov narodnogo knoziaistva, 25 maia–4 iiunia 1918 g.: Stenograficheskii otchet* (Moscow: VSNKh, 1918), 477–81.

29. *Rabochii klass v Oktiabr'skoi revoliutsii*, 455.

30. Zhuravlev, *Dekrety sovetskoi vlasti kak istoricheskii istochnik*, 268–72.

31. *Tekstil'shchik* 7–8 (25 October 1918): 15.

32. *TsGAgM*, f. 673, op. 1, d. 1125, l. 34.

33. *TsGAOR*, f. 5457, op. 2, d. 3, l. 89.

34. *Narodnoe khoziaistvo* 10 (October 1918): 72, 76.

35. *TsGAOR*, f. 5457, op. 2, d. 3, l. 25.

36. *Tekstil'shchik* 9–10 (25 December 1918): 21–22; *TsGAOR*, f. 5457, op. 2, d. 75, l. 1.

37. *Narodnoe khoziaistvo* 5 (5 July 1918): 7.

38. *Tekstil'shchik* 3–4 (12 May 1918): 17; *Tekstil'nyi rabochii* 4 (9 December 1917): 8.

39. *Natsionalizatsiia promyshlennosti v SSSR: Sbornik dokumentov i materialov, 1917–1920 gg.*, 172–75; *TsGAOR*, f. 472, op. 1, d. 7, l. 179.

40. *Natsionalizatsiia promyshlennosti v SSSR: Sbornik dokumentov i materialov, 1917–1920 gg.*, 644.

41. *TsGANKh*, f. 3338, op. 1, d. 602, l. 131. "Civil war" is obviously used here to characterize class hostility before the outbreak of full-scale civil war in May–June.

42. *TsGAOR*, f. 5457, op. 2, d. 3, l. 18.

43. *Tekstil'shchik* 7–8 (25 October 1918): 6.

44. *Tekstil'shchik* 9–10 (25 December 1918): 4–5.

45. Ibid., 11.

46. *Natsionalizatsiia promyshlennosti v SSSR: Sbornik dokumentov i materialov, 1917–1920 gg.*, 256–57.

47. *Rabochii klass: Sbornik dokumentov*, 152.

48. Ibid., 155–56.

49. *Narodnoe khoziaistvo* 1–2 (January–February 1919): 10–15.

50. *TsGAOR*, f. 5457, op. 2, d. 24, l. 1.

51. *Tekstil'shchik* 1–2 (9/22 April 1918): 17.

52. Gimpel'son, *Rabochii klass v upravlenii sovetskim gosudarstvom*, 273.

53. *TsGAOR*, f. 5457, op. 2, d. 3, ll. 40–41.

54. *Tekstil'shchik* 1–2 (9/22 April 1918): 7–8.

55. *TsGAOR*, f. 5457, op. 2, d. 3, l. 18; *Tekstil'shchik* 5–6 (1 August 1918): 14; R. M. Savitskaia, "Rabochie tovarishcheskie distsiplinarnye sudy v sovetskoi Rossii (1917–1921 gg.)," *Istoriia SSSR* 2 (March–April 1987): 136–37.

56. *Natsionalizatsiia promyshlennosti v SSSR: Sbornik dokumentov i materialov, 1917–1920 gg.*, 655.

57. Romanov, *Tekstil'shchiki Moskovskoi oblasti*, 59–60; Savitskaia, "Rabochie tovarishcheskie distsiplinarnye sudy v sovetskoi Rossii (1917–1921 gg.)," 137.

58. *TsGAOR*, f. 5457, op. 2, d. 3, l. 23; see also *Natsionalizatsiia promyshlennosti v SSSR: Sbornik dokumentov i materialov, 1917–1920 gg.*, 717.

59. *Narodnoe khoziaistvo* 8–9 (September 1918): 64.

60. *Tekstil'nyi rabochii* 7 (14/27 February 1918): 11–12. In a speech in Ivanovo–Voznesensk on January 27, M. V. Frunze, who later became People's Commissar for the Army, reiterated the idea that nationalization

was an inevitable outcome of the revolution that nevertheless required detailed preparation (*Ivanovo–Voznesenskie bol'sheviki: Sbornik dokumentov*, 174). In this usage, Moscow *Oblast'* consisted of Moscow, Tver, Iaroslavl, and Vladimir provinces. The Bolshevik party, as noted in Chapter 1, n. 1, used a different definition.

61. Laverychev, "Sozdanie organov upravleniia tekstil'noi promyshlennosti," 120–21.

62. *TsGAOR*, f. 472, op. 1, d. 7, l. 181.

63. Ibid., l. 183.

64. Ibid., l. 128.

65. Ibid., ll. 188–89.

66. Ibid., l. 187.

67. Ibid., l. 60.

68. *Tekstil'nyi rabochii* 7 (14/27 February 1918): 3; *Glavtekstil'*, 18.

69. *Protokoly 1-go s"ezda soiuzov tekstil'shchikov*, 3. My italics.

70. Ibid., 5–6.

71. Ibid., 18–19, 21–23.

72. Ibid., 5–6.

73. Ibid., 20.

74. Ibid., 34–41. Ultimately, the First Congress of the Union of Textile Workers would resolve that workers' control was but a transitional step on the road to a new type of administration in the industry. Ibid., 11, 24–30; *Narodnoe khoziaistvo* 11 (November 1918): 44.

75. *Protokoly 1-go s"ezda soiuzov tekstil'shchikov*, 63.

76. Ibid., 39.

77. Ibid., 63.

78. *Tekstil'nyi rabochii* 7 (14/27 February 1918): 7.

79. *Glavtekstil'*, 19–22; *Protokoly 1-go s"ezda soiuzov tekstil'shchikov*, 61–63; *Rabochii krai* 245 (19 September 1918): 1.

80. The Soviet state adjusted the discrepancy between the Western and Russian calendars at the beginning of February 1918. Hence, the initial Centro-Textile meeting took place February 2 (Old Style), while the second occurred three days later, on February 18 (New Style). All dates herein conform to the calendar currently in use unless otherwise noted.

81. Romanov, *Tekstil'shchiki Moskovskoi oblasti*, 51–55.

82. *Tekstil'shchik* 1–2 (9/22 April 1918): 14; Romanov, *Tekstil'shchiki Moskovskoi oblasti*, 56; Laverychev, "Sozdanie organov upravleniia tekstil'noi promyshlennosti," 125; Venediktov, *Organizatsiia gosudarstvennoi promyshlennosti v SSSR*, 286.

83. *Tekstil'shchik* 7–8 (25 October 1918): 11.

84. Romanov, *Tekstil'shchiki Moskovskoi oblasti*, 56.

85. *Narodnoe khoziaistvo* 11 (November 1918): 44.

86. *Glavtekstil'*, 23.
87. Ibid., 23–24; *TsGAOR*, f. 5457, op. 3, d. 1, l. 15; *Tekstil'shchik* 1–2 (9/22 April 1918): 9.
88. *Narodnoe khoziaistvo* 5 (5 July 1918): 28.
89. *TsGAOR*, f. 5457, op. 2, d. 7, l. 29.
90. *Natsionalizatsiia promyshlennosti v SSSR: Sbornik dokumentov i materialov, 1917–1920 gg.*, 219.
91. *TsGAOR*, f. 5457, op. 2, d. 7, l. 29.
92. *Tekstil'shchik* 5–6 (1 August 1918): 11–12.
93. *TsGAOR*, f. 5457, op. 2, d. 7, l. 101.
94. *Natsionalizatsiia promyshlennosti v SSSR: Sbornik dokumentov i materialov, 1917–1920 gg.*, 61–62.
95. G. V. Dvorkin, "Ot rabochego kontrolia k natsionalizatsii fabriki (Iz istorii 'Trekhgornoi Manufaktury')," *Uchenye zapiski Moskovskogo pedagogicheskogo instituta im. V. I. Lenina* 286 (1967): 115.
96. *Narodnoe khoziaistvo* 2 (April 1918): 44.
97. *Rabochii klass: Sbornik dokumentov*, 177–78. An arshin equals approximately 28 inches.
98. *Tekstil'shchik* 1–2 (9/22 April 1918): 5–6.
99. *TsGAOR*, f. 5457, op. 2, d. 5, ll. 10–12. The following provinces constituted the Northern Industrial Region in 1918: Petrograd, Vologda, Olonets, Archangel, Novgorod, Pskov, and the unoccupied districts of Liftiliand.
100. *Natsionalizatsiia promyshlennosti v SSSR: Sbornik dokumentovi i materialov, 1917–1920 gg.*, 329.
101. Ibid., 332.
102. Ibid., 395–96.
103. *TsGAOR*, f. 5457, op. 2, d. 75, l. 1.
104. *Natsionalizatsiia promyshlennosti v SSSR: Sbornik dokumentovi i materialov, 1917–1920 gg.*, 690–92.
105. *TsGAOR*, f. 5457, op. 2, d. 5, l. 44.
106. *Tekstil'shchik* 9–10 (25 December 1918): 3.
107. *TsGANKh*, f. 3429, op. 1, d. 162, l. 18.
108. *TsGAOR*, f. 5457, op. 2, d. 5, l. 30.
109. *Rabochii klass: Sbornik dokumentov*, 120–21.
110. *Ivanovo-Voznesenskie bol'sheviki: Sbornik dokumentov*, 180.
111. *TsGAgM*, f. 673, op. 1, d. 1125, ll. 9–11.
112. *Natsionalizatsiia promyshlennosti v SSSR: Sbornik dokumentov i materialov, 1917–1920 gg.*, 760.
113. *TsGAOR*, f. 5457, op. 2, d. 5, ll. 29, 32–33, 41.
114. Ibid., ll. 40–41.
115. *Tekstil'shchik* 5–6 (1 August 1918): 2–4.

116. *Narodnoe khoziaistvo* 11 (November 1918): 44; *TsGAOR*, f. 5457, op. 2, d. 5, l. 33.

117. *TsGAOR*, f. 5457, op. 2, d. 5, l. 66.

118. *Natsionalizatsiia promyshlennosti v SSSR: Sbornik dokumentov i materialov, 1917–1920 gg.*, 402–05, 411–15, 418–25, 439–40.

119. *TsGANKh*, f. 7659, op. 1, d. 13, l. 1.

120. *Tekstil'shchik* 3–4 (12 May 1918): 12–14.

121. *Tekstil'shchik* 5–6 (1 August 1918): 4–5.

122. *TsGAOR*, f. 5457, op. 2, d. 5, l. 1.

123. *Tekstil'shchik* 5–6 (1 August 1918): 4.

124. *Tekstil'nyi rabochii* 7 (14/27 February 1918): 5–6.

125. Ibid., 7; *TsGAOR*, f. 472, op. 1, d. 7, l. 179; *TsGAOR*, f. 472, op. 1, d. 8, ll. 1, 3.

126. *Tekstil'nyi rabochii* 4 (9 December 1917): 5.

127. T. A. Ignatenko, "Nekotorye voprosy izucheniia istorii rabochego kontrolia i natsionalizatsii promyshlennosti (1917–1918 gg.)," in *Iz istorii grazhdanskoi voiny i interventsii, 1917–1922 gg.*, 388, 392.

128. *TsGAOR*, f. 472, op. 1, d. 7, l. 185. My italics.

129. Ibid., l. 211.

130. *TsGAOR*, f. 472, op. 1, d. 8, l. 67.

131. Ignat'ev, *Moskva v pervyi god proletarskoi diktatury*, 133–34.

132. S. A. Fediukin, *Sovetskaia vlast' i burzhuaznye spetsialisty* (Moscow: Mysl', 1965), 99.

133. Drobizhev, "Bor'ba russkoi burzhuazii protiv natsionalizatsii promyshlennosti," 33–36.

134. Speaking for the Bolshevik Central Committee on the tempo of organization, Grigorii Zinoviev reported to the Sixth Congress of Soviets on November 9, 1918, that even the establishment of soviets, the institutions in whose name the party had seized office, still lay in the future in many parts of the country (*Shestoi Vserossiiskii chrezvychainyi s"ezd sovetov: Stenograficheskii otchet* [Moscow: Gosudarstvennoe izdatel'stvo, 1919], 87).

135. Drobizhev, "Obrazovanie sovetov narodnogo khoziaistva v Moskovskom promyshlennom raione," 87–91; Ignat'ev, *Moskva v pervyi god proletarskoi diktatury*, 137, 139.

136. *Tekstil'shchik* 1–2 (9/22 April 1918): 11–12; *Tekstil'nyi rabochii* 7 (14/27 February 1918): 10–11.

137. *Rabochii krai* 96 (19 March 1918): 3.

138. *TsGAOR*, f. 5457, op. 2, d. 24, l. 17. By June 25, 1918, union organization in Tver still would not have passed the stage of only passing resolutions. *TsGAOR*, f. 5457, op. 2, d. 12, l. 90.

139. *Tekstil'nyi rabochii* 5 (20 December 1917): 15–16.

140. *Tekstil'nyi rabochii* 7 (14/27 February 1918): 3.
141. *Tekstil'shchik* 7–8 (25 October 1918): 11.
142. *TsGAOR*, f. 5457, op. 2, d 7, l. 29.
143. *TsGAOR*, f. 5457, op. 2, d. 5, ll. 32, 44, 46.
144. *TsGAOR*, f. 5457, op. 2, d. 3, l. 108.
145. Laverychev, "Sozdanie organov upravleniia tekstil'noi promy-shlennosti," 138.
146. *TsGAOR*, f. 5457, op. 2, d. 3, l. 115.
147. *Tekstil'shchik* 7–8 (25 October 1918): 15.
148. *TsGAOR*, f. 5457, op. 2, d. 3, l. 90.
149. *Narodnoe khoziaistvo* 11 (November 1918): 45; Iu. K. Avadkov, *Organizatsionno-khoziaistvennaia deiatel'nost' VSNKh v organizatsii up-ravleniia sovetskoi promyshlennost'iu (1917–1921 gg.)* (Moscow: Izda-tel'stvo Moskovskogo universiteta, 1971), 113.
150. *Narodnoe khoziaistvo* 10 (October 1918): 72.
151. *TsGAgM*, f. 673, op. 1, d. 1125, ll. 13, 34.
152. O. V. Naumov, L. S. Petrosian, and A. K. Sokolov, "Kadry ruko-voditelei, spetsialistov i obsluzhivaiushchego personala promyshlennykh predpriiatii po dannym professional'noi perepisi 1918 goda," *Istoriia SSSR* 6 (November–December 1981): 98.
153. *TsGAOR*, f. 5457, op. 2, d. 24, l. 12.
154. *Rabochii klass: Sbornik dokumentov*, 288–89.
155. *Tekstil'shchik* 1–2 (9/22 April 1918): 7–8.
156. *TsGAOR*, f. 5457, op. 2, d. 7, l. 93.
157. *Tekstil'shchik* 9–10 (25 December 1918): 22–23.
158. *Tekstil'shchik* 7–8 (25 October 1918): 5.
159. *Tekstil'nyi rabochii* 7 (14/27 February 1918): 6; William J. Chase, "Moscow and Its Working Class, 1918–1928: A Social Analysis" (Ph.D. diss., Boston College, 1979), 20.
160. *Rabochii klass: Sbornik dokumentov*, 282. In this case, Moscow *Oblast'* includes the provinces Moscow, Vladimir, Voronezh, Kaluga, Kos-troma, Kursk, Nizhnii–Novgorod, Orlov, Riazan, Smolensk, Tver, Tula, and Iaroslavl.
161. Ibid., 39, 42–47.
162. *Rabochii klass: Sbornik dokumentov*, 257.
163. *Tekstil'shchik* 3–4 (12 May 1918): 18.
164. *TsGAOR*, f. 5457, op. 2, d. 5, ll. 1–5.
165. *Rabochii klass: Sbornik dokumentov*, 242–43, 246.
166. *TsGAgM*, f. 673, op. 1, d. 1125, l. 4.
167. *TsGAOR*, f. 5457, op. 2, d. 8, l. 8.
168. *Tekstil'shchik* 3–4 (12 May 1918): 11.
169. *Rabochii krai* 90 (5 March 1918): 4.

170. Chase, "Moscow and Its Working Class," 28.

171. *Natsionalizatsiia promyshlennosti v SSSR: Sbornik dokumentov i materialov, 1917–1920 gg.*, 198.

172. Ibid., 144.

173. Ibid., 220.

174. To complicate discontent, the Bolsheviks' enemies exploited long-standing urban–rural rivalries by telling peasants that the workers themselves, in instituting the eight-hour workday, caused the shortages of manufactured items. *Rabochii klass: Sbornik dokumentov*, 151, 172, 203–04; Chase, "Moscow and Its Working Class," 26.

175. *Tekstil'shchik* 5–6 (1 August 1918): 15.

176. *TsGAOR*, f. 5457, op. 2, d. 3, l. 95.

177. *Natsionalizatsiia promyshlennosti v SSSR: Sbornik dokumentov i materialov, 1917–1920 gg.*, 227–28.

178. For example, *Tekstil'shchik* 5–6 (1 August 1918): 5–6.

179. *Tekstil'shchik* 7–8 (25 October 1918): 3.

180. *Narodnoe khoziaistvo* 10 (October 1918): 31.

181. Ibid., 32.

182. *Tekstil'shchik* 9–10 (25 December 1918): 9.

183. Ibid., 9–10.

184. Ibid., 5–6.

Chapter 4

1. Mawdsley, *The Russian Civil War*, 180.

2. *Narodnoe khoziaistvo* 5 (May 1919): 61.

3. *Narodnoe khoziaistvo* 11–12 (November–December 1918): 3; *Natsionalizatsiia promyshlennosti v SSSR: Sbornik dokumentov i materialov, 1917–1920 gg.*, 587–88.

4. *Natsionalizatsiia promyshlennosti v SSSR: Sbornik dokumentov i materialov, 1917–1920 gg.*, 559–60.

5. *TsGAOR*, f. 5457, op. 3, d. 1, l. 23.

6. Drobizhev, *Glavnyi shtab sotsialisticheskoi promyshlennosti*, 188–89; E. G. Gimpel'son, *Velikii Oktiabr' i stanovlenie sovetskoi sistemy upravleniia narodnym khoziaistvom* (Moscow: Nauka, 1977), 61.

7. *Vos'moi s"ezd RKP(b), mart 1919 goda: Protokoly* (Moscow: Gosudarstvennoe izdatel'stvo politicheskoi literatury, 1959), 402–05.

8. Schapiro, *The Communist Party of the Soviet Union*, 247–48.

9. In Moscow province, for example, the number of party cells fell from 319 to 304 between May and October 1919 (*Ocherkii istorii Moskovskoi organizatsii KPSS* [Moscow: Moskovskii rabochii, 1983], 174).

10. Schapiro, *The Communist Party of the Soviet Union*, 249.

11. Timofeevskii et al., *V. I. Lenin i stroitel'stvo partii v pervye gody sovetskoi vlasti*, 142–43.

12. *TsGAOR*, f. 5457, op. 3, d. 1, l. 37.

13. *Natsionalizatsiia promyshlennosti v SSSR: Sbornik dokumentov i materialov, 1917–1920 gg.*, 472–74. To use a different measurement, of 446 textile enterprises under state direction by the end of July 1919, 428 were registered after October 1918.

14. Drobizhev, *Glavnyi shtab sotsialisticheskoi promyshlennosti*, 184.

15. *TsGAOR*, f. 5457, op. 4, d. 1, l. 10. By mid-1919, these *kusty* encompassed 460 textile enterprises.

16. *TsGAOR*, f. 5457, op. 3, d. 56, ll. 1–10, especially 6–7.

17. S. G. Strumilin, *Zarabotnaia plata i proizvoditel'nost' truda v 1913–1922 gg.* (Moscow: Voprosy truda, 1923), 56.

18. Romanov, *Tekstil'shchiki Moskovskoi oblasti*, 83, 86.

19. Quoted in Carr, *The Russian Revolution*, vol. 2, 185.

20. *Tekstil'shchik* 9–10 (25 December 1918): 18. On January 27, 1919, N. I. Lebedev recounted the growing union disenchantment with Centro-Textile before November 1918 at the union's second national congress (*TsGAOR*, f. 5457, op. 3, d. 1, l. 15).

21. As matters turned out, *VSNKh* did not confirm this part of the union resolution at the time and originally subordinated the *Glavnoe pravlenie* to Centro-Textile.

22. *TsGAOR*, f. 5457, op. 3, d. 62, ll. 6–7; *Tekstil'shchik* 9–10 (25 December 1918): 19.

23. *TsGANKh*, f. 3429, op. 1, d. 162, l. 53.

24. *TsGANKh*, f. 3338, op. 1, d. 1, l. 20.

25. *TsGANKh*, f. 3338, op. 1, d. 602, l. 132.

26. *Tekstil'shchik* 9–10 (25 December 1918): 4–5.

27. *TsGANKh*, f. 3338, op. 1, d. 602, l. 92.

28. Participants were Nogin, Matveev, Lebedev, Ganshin, and Rykunov of Glav-Textile; Asatkin, Mitrokhin, and Movshovich of the All-Russian Council of the Union of Textile Workers; E. N. Nikol'skii, administrator of the affairs of Glav-Textile; Kilevets and Kutuzov of the Bureau of the All-Russian Council; Kisel'nikov of the Ivanovo–Kineshma Union of Textile Workers; M. Rozenblium of the Moscow *Oblast'* Union; Zakharov and Kliucharev of the Melenkov Mill; V. Demikin of the Kovrov Union of Textile Workers; and V. Shetler of the Ivanovo–Voznesensk *Gubsovnarkhoz*.

29. *TsGANKh*, f. 3338, op. 1, d. 1, ll. 100–01.

30. *TsGAOR*, f. 5457, op. 3, d. 1, ll. 6, 29, 99, 128.

31. Ibid., ll. 66–69.

32. Voting delegates numbered 450: 223 Bolsheviks; 140 party sympathizers; 72 nonparty representatives; unaffiliated delegates or members of splinter groups accounted for the remainder.

33. *TsGAOR*, f. 5457, op. 4, d. 1, l. 10; *TsGAOR*, f. 5457, op. 3, d. 27, l. 1; *Tekstil'shchik* 11–12 (20 April 1919): 15.

34. *TsGANKh*, f. 3429, op. 1, d. 729, ll. 27–28; *TsGAOR*, f. 5457, op. 3, d. 29, l. 2; *SU* 12 (24 April 1919): 149–50.

35. *TsGANKh*, f. 3429, op. 1, d. 729, l. 28. Of the three versions of this document cited in footnote 34, only the copy in *TsGANKh* includes these provisions on the composition of the committee.

36. *TsGANKh*, f. 3429, op. 1, d. 717, l. 93.

37. Movshovich, from the union's Moscow branch, had served on the presidium of the First All-Russian Congress of the Union of Textile Workers and became a member of the union Central Committee in October 1919. V. E. Egorov, an additional appointee, was added to Glav-Textile in March 1919. The remaining three members were listed as Murav'ev, Timashchev, and Belov. In addition, due to Rudzutak's other responsibilities, it became necessary to replace him almost immediately with another Glav-Textile member, Asatkin. *TsGAOR*, f. 5457, op. 3, d. 27, l. 2.

38. See also *TsGAOR*, f. 5457, op. 3, d. 20, l. 4; *Tekstil'shchik* 11–12 (20 April 1919): 10.

39. In October 1919, *VSNKh* formally recognized the status Glav-Textile had informally occupied since March: it became a department of *VSNKh* with complete authority over regulation as well as raw materials, finished goods, and warehouses; it took direction from all bodies subordinate to the Central Committee of the Textile Industry; it held jurisdiction over all nationalized enterprises, whether *kust* members or not; it could alter directives on the textile industry from the local *sovnarkhozy* and other institutions; and no local government organs would be allowed to interfere with the proceedings of state textile enterprises (*TsGANKh*, f. 3338, op. 1, d. 10, l. 4).

40. *TsGAOR*, f. 5457, op. 3, d. 20, l. 4.

41. *TsGANKh*, f. 3338, op. 1, d. 10, l. 3. The assertiveness of Glav-Textile was so generally accepted in this matter that the Glav-Textile report of October 1, 1919, cited here erroneously declared that *VSNKh* had formally dissolved Centro-Textile on March 13. Another informal indicator of its rapid rise is the practice of simply crossing out "Centro-Textile" on preprinted forms and stationery. For examples, see *TsGANKh*, f. 3338, op. 1, d. 14, ll. 52, 69–70.

42. *TsGANKh*, f. 3338, op. 1, d. 598, l. 6.

43. Ibid.; *TsGANKh*, f. 3338, op. 1, d. 1, ll. 23–24.

44. *TsGAOR*, f. 5457, op. 3, d. 57, l. 21.

45. *TsGANKh*, f. 3338, op. 1, d. 598, l. 59.

46. Ibid., l. 93.

47. *Sbornik dekretov i postanovlenii po narodnomu khoziaistvu*, vol. 3, 389–90.

48. *TsGANKh*, f. 3338, op. 1, d. 598, l. 66.

49. *TsGANKh*, f. 3338, op. 1, d. 1, ll. 49–50; *TsGANKh*, f. 3338, op. 1, d. 598, ll. 67–68.

50. *TsGANKh*, f. 3338, op. 1, d. 1, l. 54.

51. Ibid., l. 60.

52. This was in keeping with a general increase in the role of the center. For example, on November 14 *VSNKh* liquidated the institutions that supervised industrial warehouses and transferred this authority to its newly created Materials Department (*TsGANKh*, f. 3429, op. 1, d. 162, ll. 45–46). On November 30, the People's Commissariat of the Interior ordered local soviets to stop interfering in the administration of nationalized factories (*Natsionalizatsiia promyshlennosti v SSSR: Sbornik dokumentov i materialov, 1917–1920 gg.*, 582–83). In Novgorod province, where local organs initiated 43 of 53 nationalizations effected in January–October 1918, a dramatic shift occurred. Central organs ordered 151 nationalizations between November 1918 and June 1919, and local institutions carried out but 24 (*Rabochii kontrol' i natsionalizatsiia promyshlennosti Novgorodskoi gubernii v 1917–1921 gg.: Sbornik dokumentov i materialov* [Novgorod: Lenizdat, 1974], 29).

53. *Narodnoe khoziaistvo* 1–2 (January–February 1919): 15–20, especially 16.

54. *TsGAOR*, f. 5457, op. 2, d. 3, l. 136.

55. *Natsionalizatsiia promyshlennosti v SSSR: Sbornik dokumentov i materialov, 1917–1920 gg.*, 449.

56. *TsGAOR*, f. 5457, op. 3, d. 1, ll. 29, 35, 37.

57. For example, *TsGANKh*, f. 3429, op. 1, d. 162, ll. 2, 13, 14; *Natsionalizatsiia promyshlennosti v SSSR: Sbornik dokumentov i materialov, 1917–1920 gg.*, 441.

58. See *TsGANKh*, f. 3429, op. 1, d. 180, l. 64.

59. *TsGANKh*, f. 3429, op. 1, d. 177, l. 3; *TsGANKh*, f. 3429, op. 1, d. 183, ll. 1–2; *Natsionalizatsiia promyshlennosti v SSSR: Sbornik dokumentov i materialov, 1917–1920 gg.*, 449.

60. In one striking example of local control, the union section in Orekhovo-Zuevo intervened directly when it decided that efficiency would better be served by combining the Vikula Morozov Factory with the Savva Morozov Factory under a single management, and both firms were simultaneously nationalized in this manner (*TsGANKh*, f. 3429, op. 1, d. 183, l. 33).

61. *TsGANKh*, f. 3429, op. 1, d. 177, ll. 76–78.

62. Ibid., ll. 45–46.

63. *Natsionalizatsiia promyshlennosti v SSSR: Sbornik dokumentov i materialov, 1917–1920 gg.*, 585.

64. For individual examples, see *TsGANKh*, f. 3429, op. 1, d. 180, ll. 4, 12, 34, 42; *TsGANKh*, f. 3429, op. 1, d. 177, l. 69.

65. *TsGANKh*, f. 3429, op. 1, d. 183, l. 16; *TsGANKh*, f. 3429, op. 1, d. 703, l. 19.

66. *Natsionalizatsiia promyshlennosti v SSSR: Sbornik dokumentov i materialov, 1917–1920 gg.*, 451, 457.

67. *TsGAOR SSL*, f. 1931, op. 3, d. 5, l. 1.

68. *TsGANKh*, f. 3429, op. 1, d. 729, ll. 1–2.

69. For an example of this process regarding a single enterprise, the E. A. Sokolikov Factory, see the proceedings of February 12–13 in *TsGANKh*, f. 3429, op. 1, d. 717, ll. 1, 12. For an action involving fourteen enterprises, see the meetings of February 5 and 10 in ibid., ll. 18–19.

70. On March 8, *VSNKh* published a composite list of 104 enterprises nationalized on eight dates in January and February. Of the total, 41 received production-group assignments (*SU* 7 (1 April 1919): 102–04).

71. *TsGAOR*, f. 5457, op. 2, d. 5, l. 46.

72. *TsGANKh*, f. 3338, op. 1, d. 1, l. 20.

73. Ibid., l. 113.

74. *TsGAOR*, f. 5457, op. 3, d. 62, l. 8; *TsGAOR SSL*, f. 1380, op. 2, d. 11, ll. 48–51. In practice, the formation of the Kineshma *kust* and the cotton-weaving factories of Kineshma *raion*, recorded January 22, followed this procedure closely (*TsGANKh*, f. 3429, op. 1, d. 729, l. 21).

75. *TsGAOR*, f. 5457, op. 3, d. 17, l. 98.

76. *Narodnoe khoziaistvo* 4 (April 1919): 42.

77. *TsGAOR*, f. 5457, op. 3, d. 1, ll. 167–68.

78. Ibid., l. 23.

79. Ibid., l. 10. Officially, the union claimed 642,518 members by January 1, 1919, although, characteristically, the document that holds this information contains mistakes in addition (*TsGAOR*, f. 5457, op. 3, d. 27, ll. 15–16).

80. *TsGAOR*, f. 5457, op. 3, d. 26, l. 10.

81. *Tekstil'shchik* 11–12 (20 April 1919): 21.

82. Ibid., l. 1.

83. *TsGAOR*, f. 5457, op. 3, d. 1, l. 6.

84. *Tekstil'shchik* 11–12 (20 April 1919): 3–4.

85. *TsGAOR*, f. 5457, op. 3, d. 13, l. 2.

86. *Ekonomicheskaia zhizn'* 51 (6 March 1919): 3.

87. These are listed in *TsGAOR*, f. 5457, op. 3, d. 11, l. 13.

88. *TsGAOR*, f. 5457, op. 3, d. 26, ll. 1–2; *TsGAOR*, f. 5457, op. 3, d. 27, ll. 10–11.

89. *TsGAOR*, f. 5457, op. 3, d. 27, l. 2.

90. *TsGAOR*, f. 5457, op. 3, d. 11, l. 3.

91. *Tekstil'shchik* 13–14 (July 1919): 15.

92. *TsGAOR*, f. 5457, op. 3, d. 11, l. 3.

93. Ibid., l. 1.

94. *Tekstil'shchik* 13–14 (July 1919): 14. When Braginskii reported on the walkout to the Third All-Russian Congress of the Union of Textile Workers (April 16–20, 1920), he told the membership that a fundamental disagreement over the organization of administrative organs had plagued the Central Committee and that Glav-Textile was not fulfilling its mandate. Consequently, "the Central Committee ran up against a series of misunderstandings, as a result of which several members walked out of the CC" (*TsGAOR*, f. 5457, op. 4, d. 1, l. 5).

95. *TsGAOR*, f. 5457, op. 3, d. 13, l. 31.

96. *TsGAOR*, f. 5457, op. 3, d. 27, ll. 15–16.

97. *TsGAOR*, f. 5457, op. 3, d. 17, l. 63; *TsGAOR*, f. 5457, op. 3, d. 28, l. 16; *TsGAOR*, f. 5457, op. 3, d. 26, ll. 21–23.

98. *TsGAOR*, f. 5457, op. 3, d. 11, l. 1.

99. *TsGAOR*, f. 5457, op. 3, d. 26, l. 11.

100. *TsGANKh*, f. 3338, op. 1, d. 1, ll. 106–07.

101. *TsGAOR*, f. 5457, op. 3, d. 27, l. 77.

102. *TsGAOR*, f. 5457, op. 3, d. 1, l. 106.

103. *TsGANKh*, f. 3338, op. 1, d. 602, l. 1.

104. *Tekstil'shchik* 11–12 (20 April 1919): 11.

105. *TsGAOR*, f. 5457, op. 3, d. 13, l. 18.

106. *Rabochii klass: Sbornik dokumentov*, 132.

107. *TsGAOR*, f. 5457, op. 3, d. 1, ll. 103–06.

108. *TsGAOR*, f. 5457, op. 3, d. 13, l. 8.

109. For a typical example, see *Ekonomicheskaia zhizn'* 62 (22 March 1919): 2.

110. *Tekstil'shchik* 11–12 (20 April 1919): 4.

111. *TsGAOR*, f. 5457, op. 3, d. 13, l. 26.

112. *Tekstil'shchik* 11–12 (20 April 1919): 15; *Tekstil'shchik* 13–14 (July 1919): 15.

113. *TsGAOR*, f. 5457, op. 3, d. 26, l. 14.

114. *TsGANKh*, f. 3338, op. 1, d. 463, l. 2.

115. *TsGANKh*, f. 3338, op. 1, d. 602, l. 15.

116. *Tekstil'shchik* 13–14 (July 1919): 8.

117. Ibid.

118. *TsGAOR*, f. 5457, op. 3, d. 28, l. 18.

119. *TsGAOR SSL*, f. 2156, op. 1, d. 1, ll. 1–24.
120. *TsGAOR*, f. 5457, op. 3, d. 57, ll. 16–19; *TsGANKh*, f. 3338, op. 1, d. 1, ll. 27–28, 65–109.
121. Examples can be found in *TsGAOR*, f. 5457, op. 3, d. 27, l. 43; *TsGAOR*, f. 5457, op. 3, d. 17, l. 69; *TsGAOR*, f. 5457, op. 3, d. 20, l. 15.
122. Iu. K. Avdakov and V. V. Borodin, *Proizvodstvennye ob "edineniia i ikh rol' v organizatsii sovetskoe promyshlennost'iu (1917–1923 gg.)* (Moscow: Izdatel'stvo Moskovskogo universiteta, 1973), 20–21.
123. *TsGAOR*, f. 5457, op. 3, d. 57, l. 1.
124. *TsGANKh*, f. 3338, op. 1, d. 463, l. 24.
125. *TsGANKh*, f. 3338, op. 1, d. 465, ll. 14–15.
126. *TsGAOR*, f. 5457, op. 3, d. 56, ll. 13–19, 25–28.
127. Bobkov, "Iz istorii organizatsii upravleniia promyshlennost'iu v pervye gody sovetskoi vlasti (1917–1920 gg.)," 131.
128. *Natsionalizatsiia promyshlennosti v SSSR: Sbornik dokumentov i materialov, 1917–1920 gg.*, 586.
129. *TsGAOR*, f. 5457, op. 3, d. 56, ll. 25–26.
130. For example, see the Iakhromo conference of December 15–16, 1918, in *TsGAOR*, f. 5457, op. 2, d. 3, ll. 136–37.
131. For a classic example, see the report of the Petrograd Kanat Factory of March 25, 1919 (*TsGAOR SSL*, f. 1380, op. 2, d. 11, l. 40). See also *TsGAOR*, f. 5457, op. 3, d. 13, l. 13.
132. *Tekstil'shchik* 11–12 (20 April 1919): 16.
133. *Tekstil'shchik* 13–14 (July 1919): 12.
134. *TsGAOR*, f. 5457, op. 3, d. 1, ll. 79–85.
135. *TsGANKh*, f. 3429, op. 1, d. 162, l. 52.
136. *TsGAOR*, f. 5457, op. 3, d. 27, l. 18; *Rabochii klass: Sbornik dokumentov*, 231–32.
137. *TsGAOR*, f. 5457, op. 3, d. 1, ll. 165–66.
138. *TsGAOR*, f. 5457, op. 3, d. 56, ll. 22–23.
139. *TsGANKh*, f. 3338, op. 1, d. 602, l. 174.
140. *TsGAOR*, f. 5457, op. 3, d. 20, l. 7.
141. *TsGAOR SSL*, f. 1847, op. 2, d. 2, l. 31.
142. *Tekstil'shchik* 11–12 (20 April 1919): 25.
143. *TsGANKh*, f. 3338, op. 1, d. 598, l. 34.
144. *TsGAOR*, f. 5457, op. 3, d. 27, ll. 77–78.
145. *TsGANKh*, f. 3338, op. 1, d. 1, ll. 31–32.
146. *TsGANKh*, f. 3338, op. 1, d. 598, l. 14.
147. *Tekstil'shchik* 13–14 (July 1919): 21.
148. For example, *TsGAOR SSL*, f. 1380, op. 2, d. 13, ll. 44, 49; *TsGAOR SSL*, f. 1380, op. 2, d. 15, ll. 447–48; *TsGANKh*, f. 3338, op. 1, d. 463, ll. 1, 4, 18.

149. *TsGANKh*, f. 3338, op. 1, d. 463, l. 1.
150. *SU* 10–11 (21 April 1919): 136.
151. *TsGAOR*, f. 5457, op. 3, d. 11, l. 2.
152. *TsGAOR*, f. 5457, op. 3, d. 1, ll. 140–41.
153. *Tekstil'shchik* 13–14 (July 1919): 11–12.
154. *TsGAOR*, f. 5457, op. 3, d. 17, ll. 61, 73, 75, 77.
155. *Tekstil'shchik* 11–12 (20 April 1919): 18.
156. Ibid., 30.
157. *TsGAOR*, f. 5457, op. 3, d. 26, l. 22.
158. *Tekstil'shchik* 13–14 (July 1919): 7.
159. Ibid., 21–22.
160. *TsGAOR*, f. 5457, op. 3, d. 26, ll. 12–13.
161. *TsGAOR*, f. 5457, op. 3, d. 27, l. 140.
162. Romanov, *Tekstil'shchiki Moskovskoi oblasti*, 86–87. It is worth noting that the textile industry enjoyed something of an advantage over other industries in its ability to utilize peat as fuel. On the other hand, working in peat bogs was both arduous and hazardous, and it had ranked among the most exploitative forms of labor in the prerevolutionary period. In the winter of 1919–1920, 30–40 percent of those assigned to peat-gathering detachments fell ill. Ibid., 112.
163. *TsGAOR*, f. 5457, op. 3, d. 57, l. 1.
164. *Tekstil'shchik* 11–12 (20 April 1919): 19, 22.
165. *SU* 28 (2 July 1919): 349–50.
166. *TsGAOR*, f. 5457, op. 3, d. 27, l. 18; *TsGAOR*, f. 5457, op. 3, d. 17, ll. 79, 81; Romanov, *Tekstil'shchiki Moskovskoi oblasti*, 66–73.
167. *TsGAOR*, f. 5457, op. 3, d. 17, l. 84.
168. *Tekstil'shchik* 11–12 (20 April 1919): 30; *Tekstil'shchik* 13–14 (July 1919): 1, 2–3.
169. *Tekstil'shchik* 13–14 (July 1919): 15.
170. *Tekstil'shchik* 11–12 (20 April 1919): 5–6; *TsGAOR*, f. 5457, op. 3, d. 20, l. 4.
171. *TsGAOR*, f. 5457, op. 3, d. 17, ll. 60, 79.
172. *Tekstil'shchik* 13–14 (July 1919): 13.
173. *TsGAOR*, f. 5457, op. 3, d. 1, l. 149.
174. *Rabochii klass: Sbornik dokumentov*, 325, 333–34.
175. *Tekstil'shchik* 13–14 (July 1919): 18.
176. *Tekstil'shchik* 11–12 (20 April 1919): 18.
177. *Rabochii klass: Sbornik dokumentov*, 334.
178. *Tekstil'shchik* 13–14 (July 1919): 15.
179. *TsGAOR SSL*, f. 1847, op. 2, d. 2, ll. 26–27.
180. Kritsman, *Geroicheskii period velikoi russkoi revoliutsii*, 94–95.
181. *SU* 1 (26 January 1919): 1–2.

182. *TsGANKh*, f. 3338, op. 1, d. 463, l. 30.
183. Ibid., l. 6.
184. *TsGAOR*, f. 5457, op. 3, d. 1, l. 100.
185. *Tekstil'shchik* 9–10 (25 December 1918): 1. The run of the initial issue numbered 10,000, which was cut to 7,000 by October 1918.
186. *TsGANKh*, f. 3338, op. 1, d. 463, l. 6.
187. *TsGAOR*, f. 5457, op. 3, d. 11, l. 5; *Tekstil'shchik* 13–14 (July 1919): 14.
188. *TsGAOR*, f. 5457, op. 3, d. 28, l. 19.

Chapter 5

1. *Vos'moi Vserossiiskii s"ezd sovetov rabochikh, krest'ianskikh, krasnoarmeiskikh i kazach'ikh deputatov: Stenograficheskii otchet (22–29 dekabria 1920 goda)* (Moscow: Gosudarstvennoe izdatel'stvo, 1921), 109–18. For an account of how little had been achieved, see the report of former Left Communist Lev Kritsman. Ibid., 28–35.
2. Ibid., 93.
3. *TsGAOR*, f. 5457, op. 5, d. 16, l. 4.
4. *TsGAOR*, f. 5457, op. 5, d. 1, l. 27.
5. For example, see the Fourth *Raion* Conference of the Klintsy Section of the Union of Textile Workers of February 20, 1920 (*TsGAOR*, f. 5457, op. 3, d. 20, ll. 41, 46).
6. *TsGAOR*, f. 5457, op. 4, d. 2, l. 10.
7. *Tekstil'shchik* 15–16 (April–May 1920): 3. Of 358 congress delegates, 286 possessed a vote and 72 had a consultative voice. Of the voting delegates, 148 were Bolsheviks, 23 were party sympathizers, 112 were listed as nonparty, 2 were Mensheviks, and 1 was a Bund member. Of the consultative delegates, 26 were Bolsheviks, 2 were sympathizers, and 44 were nonparty representatives (*TsGAOR*, f. 5457, op. 4, d. 1, l. 1).
8. Ibid., l. 2.
9. *Tekstil'shchik* 15–16 (April–May 1920): 5, 9.
10. For example, *TsGANKh*, f. 3338, op. 1, d. 598, l. 102; *TsGAOR*, f. 5457, op. 3, d. 6, l. 1; *TsGANKh*, f. 3338, op. 1, d. 465, l. 38.
11. *TsGAOR*, f. 5457, op. 3, d. 55, l. 195.
12. *7-i Vserossiiskii s"ezd sovetov rabochikh, krest'ianskikh, krasnoarmeiskikh i kazach'ikh deputatov: Stenograficheskii otchet (5–9 dekabria 1919 goda, v Moskve)* (Moscow: Gosudarstvennoe izdatel'stvo, 1920), 197.
13. Ibid.
14. Ibid., 197–98.
15. Ibid., 200.

16. Ibid., 202.
17. Ibid., 220.
18. Ibid., 223, 225.
19. Ibid., 232–33.
20. See, for example, the report of the Simbirsk Group Administration of Cloth Factories for January 1–September 1, 1919 (*TsGANKh*, f. 3338, op. 1, d. 465, l. 1).
21. *Tekstil'shchik* 17 (June 1920): 1.
22. *TsGANKh*, f. 3338, op. 1, d. 602, l. 167.
23. *TsGAOR*, f. 5457, op. 3, d. 27, l. 15.
24. These included I. I. Kutuzov, M. O. Braginskii, Gorshkov, Kisel'-nikov, Iasenev, Anisimov, Lebedev, Asatkin, and Polonskaia. The meeting made the following assignments: Kutuzov, chairman; Braginskii, secretary; Gorshkov, Rate and Norm Department; Kisel'nikov, Organizational-Instructional Department; Iasenev, Supply Department; Polonskaia, Press Department; Lebedev and Asatkin, delegates to Glav-Textile; Ammosova, Cultural–Educational Department (*TsGAOR*, f. 5457, op. 4, d. 10, l. 19; *Tekstil'shchik* 15–16 (April–May 1920): 16–18).
25. *TsGAOR*, f. 5457, op. 3, d. 29, l. 24; *Tekstil'shchik* 24 (January 1921): 27.
26. *TsGAOR*, f. 5457, op. 4, d. 3, l. 50.
27. *Tekstil'shchik* 15–16 (April–May 1920): 8.
28. *Tekstil'shchik* 20–21 (September–October 1920): 25.
29. Ibid., 4–5.
30. *TsGAOR*, f. 5457, op. 3, d. 55, l. 152.
31. Ibid., l. 158.
32. Ibid., ll. 162–68. The record does not identify the Egor'ev–Ramen chairman by name. One verst equals approximately two-thirds of a mile.
33. Ibid., l. 46.
34. *TsGAOR*, f. 5457, op. 4. d. 1, l. 44.
35. All of the protocols and records from meetings and conferences throughout 1920 contained in *TsGAOR*, f. 5457, op. 4, d. 4, demonstrate this tendency, and ll. 9, 17–24, 59, and 60–63 provide quintessential examples.
36. *TsGAOR*, f. 5457, op. 4, d. 3, l. 9.
37. *TsGAOR*, f. 5457, op. 3, d. 55, ll. 189–90.
38. *Tekstil'shchik* 23 (December 1920): 25–28.
39. *TsGAOR*, f. 5457, op. 3, d. 29, ll. 4–9, 12.
40. *TsGAOR*, f. 5457, op. 3, d. 13, l. 51.
41. *TsGAOR SSL*, f. 1380, op. 2, d. 13, l. 2.
42. *TsGAOR*, f. 5457, op. 4, d. 3, l. 35.
43. *Tekstil'shchik* 15–16 (April–May 1920): 9–10.

44. *TsGAOR*, f. 5457, op. 3, d. 27, l. 11.

45. *TsGAOR*, f. 5457, op. 3, d. 55, l. 33.

46. Ibid., ll. 193–94.

47. *TsGAOR*, f. 5457, op. 3, d. 20, ll. 28–30.

48. *7-i s″ezd sovetov*, 228–29, 239.

49. *TsGAOR*, f. 5457, op. 3, d. 13, l. 63.

50. *Deviaty s″ezd RKP(b), mart–aprel' 1920 goda: Protokoly* (Moscow: Gosudarstvennoe izdatel'stvo politicheskoi literatury, 1960), 50–56, 115–26.

51. *TsGAOR*, f. 5457, op. 4, d. 1, ll. 15, 44.

52. Ibid., l. 66.

53. *Tekstil'shchik* 22 (7 November 1920): 7.

54. *TsGANKh*, f. 3338, op. 1, d. 13, l. 1.

55. *Tekstil'shchik* 15–16 (April–May 1920): 9–10.

56. *Tekstil'shchik* 18–19 (July–August 1920): 28–29.

57. *TsGAOR*, f. 5457, op. 4, d. 52, l. 1.

58. See *TsGAOR*, f. 5457, op. 4, d. 5, ll. 36, 56.

59. For example, see the protocol of the June 14, 1920, meeting of the *VSNKh* presidium in *TsGANKh*, f. 3429, op. 1, d. 574, l. 14.

60. *TsGAOR*, f. 5457, op. 4, d. 20, l. 5.

61. *TsGANKh*, f. 3338, op. 1, d. 6, ll. 13–14.

62. Ibid., ll. 21–23, 25, 30–34.

63. *TsGAOR*, f. 5457, op. 4, d. 20, l. 6.

64. *Tekstil'shchik* 20–21 (September–October 1920): 8.

65. Avdakov and Borodin, *Proizvodstvennye ob″edineniia i ikh rol' v organizatsii sovetskoi promyshlennost'iu*, 28.

66. Kritsman, *Geroicheskii period velikoi russkoi revoliutsii*, 201.

67. See, for example, *TsGAOR*, f. 5457, op. 4, d. 5, l. 14.

68. Ibid., l. 22; for a discussion of attempts to create a national system of disciplinary courts, see Savitskaia, "Rabochie tovarishheskie distsiplinarnye sudy v Sovetskoi Rossii (1917–1921 gg.)," 136–47.

69. *TsGAOR*, f. 5457, op. 4, d. 3. l. 60.

70. *Tekstil'shchik* 24 (January 1921): 7. Italics in the original.

71. Romanov, *Tekstil'shchiki Moskovskoi oblasti*, 122.

72. See, for example, the plenary session of the Kostroma Section of June 24, 1920 (*TsGAOR*, f. 5457, op. 4, d. 5, l. 20).

73. Romanov, *Tekstil'shchiki Moskovskoi oblasti*, 123.

74. William Chase, "Voluntarism, Mobilisation and Coercion: *Subbotniki* 1919–1921," *Soviet Studies* 41 (January 1989): 125.

75. *TsGAOR*, f. 5457, op. 4, d. 5, l. 34.

76. Observation by Leopold Haimson in an address at the Russian Studies Center of Princeton University, October 31, 1983.

77. *TsGANKh*, f. 3338, op. 1, d. 598, l. 113.

78. *TsGAOR*, f. 5457, op. 4, d. 3, l. 33.

79. Romanov, *Tekstil'shchiki Moskovskoi oblasti*, 124–25.

80. *TsGAOR*, f. 5457, op. 4, d. 2, l. 13.

81. *TsGAOR*, f. 5457, op. 4, d. 5, l. 61.

82. *TsGAOR*, f. 5457, op. 4, d. 3, l. 9.

83. *Tekstil'shchik* 18–19 (July–August 1920): 13.

84. Ibid., 26. In this regard, the textile industry was in the forefront of the shock-factory experiment. *VSNKh* did not issue its formal directive on the creation of shock factories until August 26, and its Regulations appeared only on November 20.

85. Avdakov and Borodin, *Proizvodstvennye ob"edineniia i ikh rol' v organizatsii sovetskoi promyshlennost'iu*, 24.

86. *TsGAOR*, f. 5457, op. 4, d. 55, l. 8.

87. *TsGAOR*, f. 5457, op. 4, d. 54, l. 1.

88. *Tekstil'shchik* 24 (January 1921): 26.

89. Avdakov and Borodin, *Proizvodstvennye ob"edineniia i ikh rol' v organizatsii sovetskoi promyshlennost'iu*, 23.

90. On the final point, see Remington, *Building Socialism in Bolshevik Russia*, 263–64.

91. *Tekstil'shchik* 17 (June 1920): 1.

92. Ibid., 25–26.

93. *TsGAOR*, f. 5457, op. 4, d. 3, ll. 63, 86–87, 96.

94. *Tekstil'shchik* 24 (January 1921): 28.

95. *Tekstil'shchik* 20–21 (September–October 1920): 28.

96. *Tekstil'shchik* 15–16 (April–May 1920): 15.

97. William H. Chamberlin, *The Russian Revolution*, vol. 2 (New York: Grosset and Dunlap, 1935), 111.

98. *Tekstil'shchik* 15–16 (April–May 1920): 13.

99. Ibid., 15.

100. *Tekstil'shchik* 23 (December 1920): 6.

101. *TsGAOR*, f. 5457, op. 3, d. 27, l. 100.

102. *Tekstil'shchik* 15–16 (April–May 1920): 21.

103. *TsGAOR*, f. 5457, op. 3, d. 27, l. 107.

104. *TsGAOR SSL*, f. 1847, op. 2, d. 2, l. 54.

105. *Tekstil'shchik* 17 (June 1920): 25.

106. *Tekstil'shchik* 15–16 (April–May 1920): 13.

107. *TsGAOR SSL*, f. 1724, op. 1, d. 67, l. 61. To balance the example, it did prove possible for important group administrations to receive an exemption from further mobilizations, although the procedure in such a case in Nizhnii–Novgorod lasted from December 1919 to May 10, 1920 (*TsGAOR*, f. 5457, op. 4, d. 5, l. 57).

108. *TsGAOR*, f. 5457, op. 3, d. 55, l. 20.

109. *Tekstil'shchik* 25–26 (February–March 1921): 24.

110. *TsGAOR*, f. 5457, op. 4, d. 3. l. 1.

111. *Tekstil'shchik* 22 (7 November 1920): 20.

112. *TsGAOR*, f. 5457, op. 3, d. 55, l. 17.

113. *TsGAOR*, f. 5457, op. 4, d. 1, l. 68.

114. *Tekstil'shchik* 17 (June 1920): 23.

115. *TsGAOR*, f. 5457, op. 3, d. 27, l. 109.

116. *Tekstil'shchik* 17 (June 1920): 23–24.

117. *TsGAOR*, f. 5457, op. 3, d. 55, l. 39.

118. *Tekstil'shchik* 17 (June 1920): 27.

119. *Narodnoe khoziaistvo* 3–4 (February 1920): 13. For specific documentation of youths under age eighteen working eight-hour shifts, refer to the report of the Bogorodsk Section for the period September 22, 1919–January 1, 1920 (*Tekstil'shchik* 15–16 [April–May 1920]: 21).

120. *Tekstil'shchik* 18–19 (July–August 1920): 3–4.

121. *Tekstil'shchik* 20–21 (September–October 1920): 28.

122. *TsGAOR*, f. 5457, op. 3, d. 20, l. 43.

123. *Tekstil'shchik* 22 (7 November 1920): 27.

124. *TsGAOR*, f. 5457, op. 3, d. 27, ll. 25, 39, 52, 61, 101.

125. *TsGAOR*, f. 5457, op. 3, d. 17, l. 94.

126. *Tekstil'shchik* 15–16 (April–May 1920): 8.

127. *Tekstil'shchik* 23 (December 1920): 8.

128. *Tekstil'shchik* 17 (June 1920): 4.

129. *Tekstil'shchik* 20–21 (September–October 1920): 29.

130. *TsGAOR*, f. 5457, op. 3, d. 27, l. 39.

131. *Tekstil'shchik* 25–26 (February–March 1921): 4–5. The greater ability of the skilled to find work, in conjunction with the widespread closings in the industry, also help explain why the proportion of women working in textile production actually decreased between 1917 and 1919–1920.

132. *TsGANKh*, f. 3429, op. 1, d. 1347, l. 91.

133. *TsGAOR*, f. 5457, op. 4, d. 10, l. 31.

134. *TsGAOR*, f. 5457, op. 4, d. 55, l. 23.

135. *TsGAOR*, f. 5457, op. 3, d. 28, l. 6.

136. *Tekstil'shchik* 18–19 (July–August 1920): 6.

137. *TsGAOR*, f. 5457, op. 3, d. 28, l. 4.

138. *TsGAOR*, f. 5457, op. 3, d. 27, l. 59.

139. *Tekstil'shchik* 15–16 (April–May 1920): 21.

140. *TsGAOR*, f. 5457, op. 4, d. 2, ll. 14–15.

141. See, for example, *TsGAOR*, f. 5457, op. 4, d. 3, ll. 14–15; *Tekstil'shchik* 23 (December 1920): 23.

142. *TsGAOR*, f. 5457, op. 3, d. 55, l. 71.

143. *TsGAOR*, f. 5457, op. 4, d. 10, l. 1.

144. *TsGANKh*, f. 3338, op. 1, d. 13, l. 13.

145. *TsGAOR*, f. 5457, op. 3, d. 55, l. 16.

146. *Tekstil'shchik* 20–21 (September–October 1920): 10.

147. *Tekstil'shchik* 22 (7 November 1920): 27. A subsequent article listed the principal reasons for the loss of skilled workers as the front, flight to the countryside, and work in other industries (*Tekstil'shchik* 23 [December 1920]: 2).

148. *Tekstil'shchik* 18–19 (July–August 1920): 4–5; *Tekstil'shchik* 23 (December 1920): 23; *TsGAOR*, f. 5457, op. 4, d. 5, l. 35.

149. *Tekstil'shchik* 15–16 (April–May 1920): 11–12.

150. *TsGAOR*, f. 5457, op. 3, d. 27, l. 74.

151. *Tekstil'shchik* 17 (June 1920): 22.

152. *TsGAOR*, f. 5457, op. 4, d. 3. l. 51.

153. *Tekstil'shchik* 15–16 (April–May 1920): 20.

154. *TsGAOR*, f. 5457, op. 4, d. 3, l. 61.

155. *TsGAOR*, f. 5457, op. 3, d. 17, l. 94.

156. *Tekstil'shchik* 18–19 (July–August 1920): 19.

157. *TsGAOR*, f. 5457, op. 4, d. 3, l. 9.

158. *Tekstil'shchik* 22 (7 November 1920): 10.

159. *Tekstil'shchik* 17 (June 1920): 24–25.

160. *Tekstil'shchik* 15–16 (April–May 1920): 16.

161. *TsGAOR*, f. 5457, op. 4, d. 3, l. 26.

162. Multiple examples could be produced, but see especially *TsGAOR*, f. 5457, op. 5, d. 5, l. 2; d. 14, l. 3; d. 16, ll. 1, 3–4; d. 31, l. 1; d. 34, l. 1; and d. 35a, ll. 1–2.

Chapter 6

1. In this regard, what we have seen of the career of Asatkin serves as something of a metaphor for the experiences of the leadership of the textile union in 1917–1921. Beginning as a champion of local autonomy in Iva-novo-Kineshma in 1917, he became an outspoken and consistent proponent of the centralization of authority within the union as a member of its national leadership during the civil war. By 1920–1921, he also emerged as an outspoken participant in the Workers' Opposition, advocating union maneuverability against superordinate state and party organs. Asatkin's politics demonstrate one important insight about the Workers' Opposition—that it was in this instance more the advocate for the independence of union leaders than for rank-and-file democracy by the members. More to

the point, the case of Asatkin within the leadership of the textile union illustrates the isolation of the working-class organs caught between the highest levels of policymaking and the demands of the grass roots.

2. See Chapter 1, n. 5.

3. For fuller treatments, see Malle, *The Economic Organization of War Communism*, passim; Richard Sakwa, "The Commune State in Moscow in 1918," *Slavic Review* 46 (Fall/Winter 1987): 429–49.

4. For a discussion of this point, refer to the remarks of William Rosenberg and responses by Vladimir Brovkin and Moshe Lewin in *Slavic Review* 44 (Summer 1985).

Select Bibliography

Archives

Central State Archive of the October Revolution (TsGAOR)

fond 472 Central Council of Factory Committees
fond 5457 Union of Textile Workers

Central State Archive of the National Economy (TsGANKh)

fond 3338 Glav-Textile
fond 3429 Supreme Council of the National Economy

Central State Archive of the City of Moscow (TsGAgM)

fond 673 N. N. Konshin Factory

Central State Archive of the October Revolution and Building
    of Socialism in Leningrad (TsGAOR SSL)

fond 1380 Kanat Factory
fond 1559 Petr Anisimov
fond 1724 Management of the Petrograd Wool Factories
fond 1847 Bor'ba Factory
fond 1916 Krasnaia Znamia Factory
fond 1931 First Nevskii Spinning Factory
fond 1953 V. P. Nogin
fond 2156 Management of the Flax–Hemp–Jute Factories of the *Sovnar-khoz* of the Northern Industrial Region

Hoover Institution on War, Revolution, and Peace
Jay K. Zawodny file

Published Documents and Primary Sources

Arskii, R. [A. Radzushevskii]. *Regulirovanie promyshlennosti*. Moscow: VSNKh, 1919.

Bernshtam, M. S., ed. *Nezavisimoe rabochee dvizhenie v 1918 godu: Dokumenty i materialy*. Paris: YMCA Press, 1981.

Dekrety Sovetskoi vlasti, 1917–1920. 9 vols. Moscow: Gosudarstvennoe izdatel'stvo politicheskoi literatury, 1957–1978.

Deviaty s"ezd RKP(b), mart–aprel' 1920 goda: Protokoly. Moscow: Gosudarstvennoe izdatel'stvo politicheskoi literatury, 1960.

Desiaty s"ezd RKP(b), mart 1921 goda: Stenograficheskii otchet. Moscow: Gosudarstvennoe izdatel'stvo politicheskoi literatury, 1963.

Ekonomicheskaia zhizn' SSSR: khronika sobytii i faktov, 1917–1965, vol. 1. Moscow: Izdatel'stvo Sovetskaia entsiklopediia, 1967.

Ekonomicheskoe polozhenie Rossii nakanune Velikoi Oktiabr'skoi Sotsialisticheskoi revoliutsii: Dokumenty i materialy. 2 vols. Moscow: Izdatel'stvo Akademii nauk SSSR, 1957.

Estafeta pokolenii: Sbornik dokumentov i materialov. Iaroslavl: [n.p.], 1965.

Fabrichno-zavodskaia promyshlennost' v period 1913–1918 gg.: Vserossiiskaia promyshlennsia i professional'naia perepis' 1918 g. Vyp. 1 i 2. Moscow: Tsentral'noe statisticheskoe upravlenie, 1926.

Glavnoe upravlenie tekstil'noi promyshlennosti: Kratkii otchet glavnogo pravlenii tekstil'nykh predpriiatii RSFSR. Moscow: [n.p.], 1920.

Gurovicha, A. "Vysshi Sovet Narodnogo Khoziaistva." *Arkhiv Russkoi revoliutsii* 6 (1922): 304–31.

Ivanovo-Voznesenskie bol'sheviki v period podgotovki i provedeniia Velikoi Oktiabr'skoi revoliutsii: Sbornik dokumentov. Ivanovo: Ivanskoe oblastnoe gosudarstvennoe izdatel'stvo, 1947.

KPSS v rezoliutsiiakh i resheniiakh s"ezdov, konferentsii i plenumov TsK, tom pervyi, 1898–1917. Moscow: Gosudarstvennoe izdatel'stvo politicheskoi literatury, 1970.

Krumin, G. I. *Organizatsiia i upravlenie proizvodstvom*. Moscow: 10-aia Gostip, 1920.

Larin, Iu. [M. A. Lur'e]. *Proizvodstvennaia propaganda i sovetskoe khoziaistvo na rubezhe 4-go goda*. Moscow: Gosudarstvennoe izdatel'stvo, 1920.

Larin, Iu., and L. Kritsman. *Ocherk khoziaistvennoi zhizni i organizatsiia narodnogo khoziaistva Sovetskoi Rossii (1 noiabria 1917–1 iiulia 1920)*. Moscow: Gosudarstvennoe izdatel'stvo, 1920.

Lenin, Vladimir I. *Polnoe sobranie sochineniia*. 5th ed. Moscow: Gosudarstvennoe izdatel'stvo politicheskoi literatury, 1962.

Materialy po istorii SSSR. Vol. 3, Rabochii kontrol' i natsionalizatsiia krupnoi promyshlennosti v Ivanovo-Voznesenskoi gubernii. Moscow: Izdatel'stvo Akademii nauk SSSR, 1956.

Miliutin, V. P. *Narodnoe khoziaistvo Sovetskoi Rossii: Kratkii ocherk organizatsii i upravleniia i polozheniia promyshlennosti Sovetskoi Rossii*. Moscow: VSNKh, 1920.

Moskva, Oktiabr', Revoliutsiia: Dokumenty i vospominaniia. Moscow: Moskovskii rabochii, 1987.

Natsionalizatsiia promyshlennosti i organizatsiia sotsialisticheskogo proizvodstva v Petrograde. 2 vols. Leningrad: Izdatel'stvo Leningradskogo universiteta, 1958–1960.

Natsionalizatsiia promyshlennosti v SSSR: Sbornik dokumentov i materialov, 1917–1920 gg. Moscow: Gosudarstvennoe izdatel'stvo politicheskoi literatury, 1954.

Osinskii, N. [V. V. Obolenskii]. *Stroitel'stva sotsializma*. Moscow: Kommunist, 1918.

Piataia Vserossiiskaia konferentsiia professional'nykh souizov (3–7 noiabria 1920 g.): Stenograficheskii otchet. Moscow: [n.p.], 1921.

Profsoiuzy SSSR: Dokumenty i materialy, vol. 1 (1905–1917 gg.). Moscow: Profizdat, 1963.

Protokoly 4-oi konferentsii fabrichno-zavodskikh komitetov i professional'nykh soiuzov g. Moskvy. Moscow: VTsSPS, 1919.

Protokoly I-go Vserossiiskogo s"ezda professional'nykh soiuzov tekstil'shchikov i fabrichnykh komitetov. Moscow: Izdanie Vserossiiskogo soveta professional'nykh soizov tekstil'shchikov, 1918.

Protokoly shestogo s"ezda RSDRP(b). Moscow: Partiinoe izdatel'stvo, 1934.

Rabochii klass Sovetskoi Rossii v pervyi god diktatury proletariata: Sbornik dokumentov i materialov. Moscow: Nauka, 1964.

Rabochii kontrol' i natsionalizatsiia promyshlennosti Novgorodskoi gubernii v 1917–1921 gg.: Sbornik dokumentov i materialov. Novgorod: Lenizdat, 1974.

Resheniia partii i pravitel'stva po khoziaistvennym voprosam, vol. 1 (1917–1928 gody). Moscow: Gosudarstvennoe izdatel'stvo politicheskoi literatury, 1967.

Revoliutsionnoe dvizhenie v Rossii nakanune Oktiabr'skogo vooruzhennogo vosstaniia (1–24 oktiabria 1917 g.). Moscow: Izdatel'stvo Akademii nauk SSSR, 1962.

Rezoliutsii pervogo Vserossiiskogo s"ezda sovetov narodnogo khoziaistva. Moscow: VSNKh, 1918.

Rezoliutsii vtorogo Vserossiiskogo s"ezda sovetov narodnogo khoziaistva. Moscow: VSNKh, 1919.

Sbornik dekretov i postanovlenii po narodnomu khoziaistvu. 3 vols. Moscow: [n.p.], 1918–1921.

Sbornik zhurnala "Narodnoe khoziaistvo," kniga pervaia. Moscow: VSNKh, 1918.

Sed'maia (aprel'skaia) Vserossiiskaia konferentsiia RSDRP (bol'shevikov): Protokoly. Moscow: Gosudarstvennoe izdatel'stvo politicheskoi literatury, 1958.

Sed'moi ekstrennyi s"ezd RKP(b): Stenograficheskii otchet. Moscow: Gosudarstvennoe izdatel'stvo politicheskoi literatury, 1962.

7-i Vserossiiskii s"ezd sovetov rabochikh, krest'ianskikh, krasnoarmeiskikh i kazach'ikh deputatov: Stenograficheskii otchet (5–9 dekabria 1919 goda, v Moskve). Moscow: Gosudarstvennoe izdatel'stvo, 1920.

Shestoi s"ezd RSDRP (bol'shevikov): Avgust 1917 goda. Protokoly. Moscow: Gosudarstvennoe izdatel'stvo politicheskoi literatury, 1958.

Smit, M. N. *Ocherki perekhodnogo perioda.* Moscow: SEC, 1920.

Stepanov [Stepanov-Skvortsov], I. *Ot rabochemu kontrolia k rabochemu upravleniiu.* Moscow: Zhizn' i znanie, 1918.

Syromolotov, F. F. *Finansirovanie natsionalizirovannoi promyshlennosti po VSNKh v 1919 g.* Moscow: VSNKh, 1921.

Trudy I Vserossiiskii s"ezd sovetov narodnogo khoziaistva, 25 maia–4 iiunia 1918 g.: Stenograficheskii otchet. Moscow: VSNKh, 1918.

Trudy pervago ekonomicheskago s"ezda Moskovskago promyshlennago raiona. Moscow: Izdanie Moskovskago raionnago ekonomicheskago komiteta, 1918.

Uprochenie sovetskoi vlasti v Moskve i Moskovskoi gubernii: Dokumenty i materialy. Moscow: Moskovskii rabochii, 1958.

V gody grazhdanskoi voiny. Ivanovo-Voznesenskie bol'sheviki v period inostrannoi voennoi interventsii i grazhdanskoi voiny. Sbornik dokumentov i materialov. Ivanovo: Ivanskoe knizhnoe izdatel'stvo, 1957.

Vos'moi s"ezd RKP(b), mart 1919 goda: Protokoly. Moscow: Gosudarstvennoe izdatel'stvo politicheskoi literatury, 1959.

Vos'moi Vserossiiskii s"ezd sovetov rabochikh, krest'ianskikh, krasnoarmeiskikh i kazach'ikh deputatov: Stenograficheskii otchet (22–29 dekabria 1920 goda). Moscow: Gosudarstvennoe izdatel'stvo, 1921.

Za vlast' sovetov: Sbornik dokumentov i vospominanii. Ivanovo: Verkhne-Volzhskoe knizhnoe izdatel'stvo, 1967.

Memoirs and Factory Histories

Abashkina, M. A. et al. *Povest' o trekh.* Moscow: Profizdat, 1935.

Andrianov, V. I., and V. V. Solov'ev. *Gavrilov-Iamskie tkachi.* Iaroslavl: Iaroslavskoe knizhnoe izdatel'stvo, 1963.

Babichev, V. A. *Fabrika "Krasnaia talka."* Ivanovo: Ivanskoe knizhnoe izdatel'stvo, 1953.

Babichev, V. A., I. I. Zimin, and V. M. Smirnov. *Teikovskii khlopchatobumazhnyi: Istoricheskii ocherk.* Iaroslavl: Verkhne-Volzhskoe knizhnoe izdatel'stvo, 1966.

Babichev, V. A., and I. D. Zolkin. *Fabrika imeni rabochego Fëdor Zinov'-eva.* Ivanovo: Ivanskoe knizhnoe izdatel'stvo, 1956.

Braginskii, M. O. *Nasha pobeda.* Moscow: Izdatel'stvo Ts. K. Soiuza tekstil'shchikov, 1925.

Gorelkin, G. "Zhizn' i rabota fabriki Ivana Garelina v 1917 goda (po vospominaniiam uchastnikov)." *Na Leninskom puti* 12 (1926): 62-71.

Istoriia vozniknoveniia i razvitiia Gorodishchenskoi sukonnoi fabriki (po vospominaniiam S. I. Chertverikov). Moscow: Izdatel'stvo T-va Riabushinskikh, 1918.

Klimokhin, S. K. *Kratkaia istoriia stachki tekstil'shchikov Ivanovo-Kineshma promyshlennoi oblasti (s 21-go oktiabria po 17-e noiabria 1917 g.).* Kineshma: Tipografiia khoziastv. sektsiia SRSiKD, 1918.

Korolev, G. K. *Ivanovo-Kineshemskie tekstil'shchiki v 1917 godu (iz vospominanii tekstil'shchika).* Moscow: Izdatel'stvo VTsSPS, 1927.

Kozhukov, A. M. "Vospominaniia rabochego Tsindelevskoi manufaktury." In *Moskovskie Bol'sheviki v ogne revoliutsionnykh boev,* 270-82. Moscow: Mysl', 1976.

Krasnaia presnia: Ocherki po istorii zavodov. Moscow: Moskovskoe tovarishchestvo pisatelei, 1934.

Kurakhtanov, V. M. *Pervaia sittsenabivaniia.* Moscow: Sotsial'no-ekonomicheskoi literatury, 1960.

Lapitskaia, S. *Byt rabochikh Trekhgornoi manufaktury.* Moscow: Istoriia zavodov, 1935.

Leshukov, T. N. *Poltora v stroiu.* Ivanovo: Ivanskoe knizhnoe izdatel'stvo, 1962.

Nogin, V. P. *Fabrika Palia.* Leningrad: Tip. Pvorka, 1924.

Petrova, V. "Likinskie tekstil'shchiki." In *Za vlast' sovetov,* 417-24. Moscow: Moskovskii rabochii, 1957.

Pobeda Velikoi Oktiabr'skoi sotsialisticheskoi revoliutsii: Sbornik vospominanii. Moscow: Izdatel'stvo politicheskoi literatury, 1958.

Rabotnitsa na sotsialisticheskoi stroike: Sbornik avtobiografii rabotnits.
Moscow: Partiinoe izdatel'stvo, 1932.
Sokolov, V. M. *Fabrika imeni O. A. Varentsovoi.* Ivanovo: Ivanskoe
knizhnoe izdatel'stvo, 1955.
———. *Fabrika imeni S. I. Balashova.* Ivanovo: Ivanskoe knizhnoe izda-
tel'stvo, 1961.
Staraia i novaia Danilovka: Rasskazy rabochikh f-ki im. M. V. Frunze.
Moscow: Moskovskii rabochii, 1940.
Tsvetkov, G. K., P. G. Khlopotukhin, and P. G. Andreev. *Iartsevo: Ocherki
po istorii rabochego klassa i revoliutsionnogo dvizhenniia na Iart-
sevskoi fabrike.* Moscow: Partizdat, 1932.

Periodicals and Newspapers

Ekonomicheskaia zhizn'. 1918–1920.
Kommunist. 1918.
Narodnoe khoziaistvo. 1918–1921.
Novyi put'. 1917–1918.
Pravda. 1918–1921.
Rabochii krai. 1918–1920.
*Sobranie uzakonenii i rasporiazhenii rabochego i krest'ianskogo pravi-
tel'stva.* 1917–1920.
Sotsial-Demokrat. 1917.
Tekstil'shchik. 1918–1921.
Tekstil'nyi rabochii. 1917–1918.
Utro Rossii. 1917.

Unpublished Works

Avrich, Paul. "The Russian Revolution and the Factory Committees."
Ph.D. diss., Columbia University, 1961.
Bobkov, K. I. "Sotsialisticheskoe obobshchestvlenie krupnoi tekstil'noi
promyshlennosti (1917–1920 gg.)." Candidate diss., Moscow State
University, 1957.
Buchanan, Herbert Ray. "Soviet Economic Policy For the Transition Pe-
riod: The Supreme Council of the National Economy, 1917–1920."
Ph.D. diss., Indiana University, 1972.
Chase, William J. "Moscow and Its Working Class, 1918–1928: A Social
Analysis." Ph.D. diss., Boston College, 1979.

Guroff, Gregory. "The State and Industrialization in Russian Economic Thought." Ph.D. diss., Princeton University, 1970.

Holman, Glenn P. "'War Communism,' or the Beseiger Beseiged: A Study of Lenin's Political and Social Objectives from 1918 to 1921." Ph.D. diss., Georgetown University, 1974.

Iurasov, I. N. "Rabochii kontrol' v promyshlennosti Ivanovo-Voznesenskogo raiona (gubernii) v 1917–1918 gg." Candidate diss., Leningrad State Pedagogical Institute, 1951.

Joffe, Muriel. "The Cotton Manufacturers in the Central Industrial Region, 1880s–1914: Merchants, Economics, and Politics." Ph.D. diss., University of Pennsylvania, 1981.

Lih, Lars T. "Bread and Authority in Russia: Food Supply and Revolutionary Politics, 1914–1921." Ph.D. diss., Princeton University, 1983.

Sirianni, Carmen J. "Workers' Control and Socialist Democracy: The Early Soviet Experience in Comparative Perspective." Ph.D. diss., SUNY-Binghamton, 1979.

Sozinov, E. M. "Kommunisticheskaia partiia v bor'be za perekhod k rabochemu upravleniiu i natsionalizatsii krupnoi promyshlennosti." Candidate diss., Moscow State University, 1953.

Selected Secondary Sources in Russian

Alekseev, G. P., and E. Ivanov. *Profsoiuzy v period stroitel'stva kommunizma.* Moscow: Profizdat, 1968.

Aleshchenko, N. M. "Moskovskii sovet v 1918–1920 gg." *Istoricheskie zapiski* 91 (1973): 81–112.

Ananov, I. N. *Razvitie organizatsionnykh form upravleniia promyshlennost'iu SSSR.* Moscow: Gosiurizdat, 1958.

Ankudinova, L. E. *Natsionalizatsiia promyshlennosti v SSSR (1917–1920 gg.).* Leningrad: Izdatel'stvo Leningradskogo universiteta, 1963.

Astaf'ev, A. "Stachka tkachei Ivanovo-Voznesenskoi manufaktury v 1895 g." *Krasnyi arkhiv* 5 (1935): 110–19.

Avdakov, Iu. K. *Organizatsionno-khoziaistvennaia deiatel'nost' VSNKh v organizatsii upravleniia sovetskoi promyshlennost'iu (1917–1921 gg.).* Moscow: Izdatel'stvo Moskovskogo universiteta, 1971.

Avdakov, Iu. K., and V. V. Borodin. *Proizvodstvennye ob"edineniia i ikh rol' v organizatsii sovetskoi promyshlennost'iu (1917–1932 gg.).* Moscow: Izdatel'stvo Moskovskogo universiteta, 1973.

Baevskii, D. A. "Rol sovnarkhozov i profsoiuzov v organizatsii sotsialisticheskogo promyshlennogo proizvodstva v 1917–1920 gg." *Istoricheskie zapiski* 64 (1959): 3–46.

Berkhin, I. B. *Ekonomicheskaia politika sovetskogo gosudarstva v pervye gody sovetskoi vlasti.* Moscow: Nauka, 1970.

———. *Voprosy istorii perioda grazhdanskoi voiny (1918–1920 gg.) v sochineniiakh V. I. Lenina.* Moscow: Nauka, 1981.

Bobkov, K. I. "Iz istorii organizatsii upravleniia promyshlennost'iu v pervye gody sovetskoi vlasti (1917–1920 gg.) (Na materialakh tekstil'noi promyshlennosti)." *Voprosy istorii* 4 (April 1957): 119–32.

Bugai, N. F. "Chrezvychainye organy sovetskoi vlasti i vosstanovlenii promyshlennosti i transporta (1918–1921 gg.)." *Istoricheskie zapiski* 112 (1985): 31–70.

Burdzhalov, E. N. *Vtoraia russkaia revoliutsiia, vol. 2. Moskva. Front. Periferiia.* Moscow: Nauka, 1971.

Drobizhev, V. Z. "Bor'ba russkoi burzhuazii protiv natsionalizatsii promyshlennosti v 1917–1920 gg.," *Istoricheskie zapiski* 68 (1961): 28–50.

———. "Sotsialisticheskoe obobshchestvlenie promyshlennosti v SSSR." *Voprosy istorii* 6 (June 1964): 43–64.

———. *Glavnyi shtab sotsialisticheskoi promyshlennosti: Ocherkii istorii VSNKh, 1917–1932.* Moscow: Mysl', 1966.

———. *Krasnogvardeiskaia ataka na kapital.* Moscow: Izdatel'stvo politicheskoi literatury, 1976.

Dvorkin, G. V. "Ot rabochego kontrolia k natsionalizatsii fabrik (Iz istorii 'Trekhgornoi manufaktury')." *Uchenye zapiski Moskovskogo pedagogicheskogo instituta im. V. I. Lenina* 286 (1967): 101–28.

Fediukin, S. A. *Sovetskaia vlast' i burzhuaznye spetsialisty.* Moscow: Mysl', 1965.

Fëdorov, K. G. *VTsIK v pervye gody sovetskoi vlasti, 1917–1920 gg.* Moscow: Gosudarstvennoe izdatel'stvo iuridicheskoi literatury, 1957.

Fëdorov, V. I. "Ivanovo–Voznesenskie rabochie na zashchite zavoevanii Velikogo Oktiabria (oktiabr' 1917 g.–1920 g.)." *Istoricheskie zapiski* 100 (1977): 334–54.

Gaponenko, L. S. *Rabochii klass Rossii v 1917 godu.* Moscow: Nauka, 1970.

Gavrilova, I. N. "Demograficheskaia istoriia Moskvy v pervoe desiatiletie sovetskoi vlasti." *Vestnik Moskovskogo universiteta, Seriia 8, Istoriia* 3 (1983): 42–50.

Gimpel'son, E. G. *"Voennyi kommunizm": Politika, praktika, ideologiia.* Moscow: Mysl', 1973.

———. *Sovetskii rabochii klass.* Moscow: Nauka, 1974.

———. *Velikii Oktiabr' i stanovlenie sovetskoi sistemy upravleniia narodnym khoziaistvom. Moscow: Nauka, 1977.*

———. "*Rabochii klass v upravlenii sovetskim gosudarstvom (noiabr' 1917–1920 gg.*)." *Voprosy istorii* 11 (November 1981): 25–40.

———. *Rabochii klass v upravlenii sovetskim gosudarstvom noiabr' 1917– 1920 gg.* Moscow: Nauka, 1982.

———. "Bor'ba za formirovanie soznatel'noi trudovoi distsipliny v pervye gody sovetskoi vlasti." *Istoriia SSSR* 6 (November–December 1983): 114–25.

Gladkov, I. A. *Ocherki stroitel'stva sovetskogo plannogo khoziaistva v 1917–1918 gg.* Moscow: Gosudarstvennoe izdatel'stvo politicheskoi literatury, 1950.

———. *Ocherki sovetskoi ekonomiki, 1917–1920 gg.* Moscow: Gosudarstvennoe izdatel'stvo politicheskoi literatury, 1956.

———. *V. I. Lenin—Organizator sotsialisticheskoi ekonomiki.* Moscow: Gosudarstvennoe izdatel'stvo politicheskoi literatury, 1960.

Gorodetskii, E. N. *Rozhdenie sovetskogo gosudarstva.* Moscow: Nauka, 1965.

Grunt, A. Ia. *Moskva 1917-i: Revoliutsiia i kontrrevoliutsiia.* Moscow: Nauka, 1976.

Iasev, G. S. *Rol' tekstil'noi promyshlennosti v genezise i razvitii kapitalizma v Rossii, 1760–1860.* Leningrad: Nauka, 1970.

Ignatenko, T. A. *Sovetskaia istoriografiia rabochego kontrolia i natsionalizatsii promyshlennosti v SSSR, 1917–1967.* Moscow: Nauka, 1971.

Ignat'ev, G. S. *Moskva v pervyi god proletarskoi diktatury.* Moscow: Nauka, 1975.

Ionov, I. N. *Profsoiuzy rabochikh Moskvy v revoliutsii 1905–1907 gg.* Moscow: Nauka, 1986.

Istoriia rabochikh Moskvy, 1917–1945. Moscow: Nauka, 1983.

Khesin, S. S. *Stanovlenie proletarskoi diktatury v Rossii.* Moscow: Nauka, 1975.

Kim, M. P., ed. *Istoriografiia sotsialisticheskogo i kommunisticheskogo stroitel'stva v SSSR.* Moscow: Nauka, 1962.

———. *Voprosy istoriografii rabochego klassa SSSR.* Moscow: Mysl', 1970.

Korneev, A. M. *Tekstil'naia promyshlennost' SSSR i puti ee razvitiia.* Moscow: Izdatel'stvo literatury po legkoi promyshlennosti, 1957.

Kostina, R. V. "Moskovskii gorodskoi sovnarkhoz v reshenii voprosov upravleniia promyshlennost'iu stolitsy (1918–1920 gg.)." *Istoriia SSSR* 3 (May–June 1984): 116–27.

Kritsman, L. N. *Geroicheskii period velikoi russkoi revoliutsii.* Moscow: Gosudarstvennoe izdatel'stvo, 1924.

Langovoi, N. P. et al. *Fabrichno-zavodskaia promyshlennost' i torgovlia Rossii.* St. Petersburg: Izdanie Departamenta Torgovli i Manufaktur Ministerstva Finansov, 1893.

Laverychev, V. Ia. "Protsess monopolizatsii khlopchatobumazhnoi promyshlennosti Rossii (1900–1914 gg.)." *Voprosy istorii* 2 (1960): 137–51.

———. *Monopolisticheskoi kapital v tekstil'noi promyshlennosti Rossii, 1900–1917 gg.* Moscow: Izdatel'stvo Moskovskogo universiteta, 1963.

Laverychev, V. Ia., and A. M. Solov'eva. *Boevoi pochin rossiiskogo proletariata: K 100-letiiu Morozovskoi stachki 1885 g.* Moscow: Mysl', 1985.

Lavrin, V. A. *Vozniknovenie revoliutsionnoi situatsii v Rossii v gody pervoi mirovoi voiny.* Moscow: Izdatel'stvo Moskovskogo universiteta, 1984.

Lisetskii, A. M. *Bol'sheviki vo glave massovykh stachek.* Kishinev: Izdatel'stvo Shtiintsa, 1974.

"Materialy k istorii rabochego kontrolia nad proizvodstvom (1917–1918 gg.)." *Krasnyi arkhiv* 6 (1940): 106–29.

Meshalin, I. V. *Tekstil'naia promyshlennost' krest'ian Moskovskoi gubernii v XVIII i pervoi polovine XIX veka.* Moscow: Izdatel'stvo Akademii nauk SSSR, 1950.

Mikhail'kov, R. P. *Ocherki istorii Trekhgornoi manufaktury v sviazi s istoriei tekstil'noi promyshlennosti za 170 let.* Rostov-on-Don: [n.p.], 1972.

Miliutin, V. P. *Istoriia ekonomicheskogo razvitiia SSSR, 1917–1927.* Moscow: Gosizdat, 1928.

Mints, I. I., ed. *Velikii Oktiabr' i zashchita ego zavoevanii: Pobeda sotsialisticheskoi revoliutsii.* Moscow: Nauka, 1987.

———. *Velikii Oktiabr' i zaschchita ego zavoevanii: Zashchita sotsialisticheskogo Otechestva.* Moscow: Nauka, 1987.

Mints, I. I. et al., eds. *Iz istorii grazhdanskoi voiny i interventsii, 1917–1922 gg.* Moscow: Nauka, 1974.

Molodtsygin, M. A. *Raboche-krest'ianskii soiuz, 1918–1920.* Moscow: Nauka, 1987.

Nikolaev, P. A. *Rabochie-metallisty teesntral'no-promyshlennogo raiona Rossii v bor'be za pobedu Oktiabr'skoi revoliutsii (mart–noiabr' 1917 g.).* Moscow: Izdatel'stvo VPSH i AON pri TsK KPSS, 1960.

Nosach, V. I. "Pervye shagi k novoi distsipline truda (oktiabr' 1917–1920 gg.)." *Voprosy istorii* 1 (January 1984): 68–79.

Ocherki po istorii revoliutsionnogo dvizheniia i Bol'shevistskoi organizatsii v Baumanskoi raione. Moscow-Leningrad: Moskovskii rabochii, 1928.

Oktiabr'skaia revoliutsiia v Nizhegorodskoi gubernii. Nizhnii Novgorod: Nizhegorodskii istpartotdel gubkoma VPK(b), 1927.

Pankratova, A. *Fabzavkomy Rossii v bor'be za sotsialisticheskuiu fabriku.* Moscow: Izdatel'stvo "Krasnaia nov'," 1923.

Pazhitnov, K. A. *Ocherkii istorii tekstil'noi promyshlennosti dorevoliutsionnoi Rossii: Sherstianaia promyshlennost'.* Moscow: Izdatel'stvo Akademii nauk SSSR, 1958.

Perazich, V. D. *Tekstili Leningrada v 1917 g.* Leningrad: Izdanie Leningradskogo gubotdela vsesoiuznogo soiuza tekstil'shchikov, 1927.

Pogudin, V. I. *Osushchestvlenie Leninskogo plana stroitel'stvo sotsializma v SSSR.* Moscow: Znanie, 1969.

Polikarpov, V. D. *Prolog grazhdanskoi voiny v Rossii.* Moscow: Nauka, 1976.

Rabochii klass i rabochee dvizhenie v Rossii v 1917 g. Moscow: Nauka, 1964.

Rabochii klass v Oktiabr'skoi revoliutsii i na zashchite ee zavoevanii, 1917–1920 gg. Vol. 1. Moscow: Nauka, 1984.

Rashin, A. G. *Formirovanie rabochego klassa Rossii: Istoriko-ekonomicheskie ocherki.* Moscow: Izdatel'stvo sotsial'no-ekonomicheskoi literatury, 1958.

Razumovich, N. N. *Organizatsionno-pravovye formy sotsialisticheskogo obobshchestvleniia promyshlennosti v SSSR, 1917–1920 gg.* Moscow: Izdatel'stvo Akademii nauk SSSR, 1959.

Romanov, F. A. *Tekstil'shchiki Moskovskoi oblasti v gody grazhdanskoi voiny.* Moscow: Profizdat, 1939.

Rozenfel'd, Ia. S. *Promyshlennaia politika SSSR, 1917–1925 gg.* Moscow: Gosplan, 1926.

Savel'ev, Iu. S. *V pervyi god Velikogo Oktiabria.* Moscow, Mysl', 1985.

Savitskaia, R. M. "Rabochie tovarishcheskie distsiplinarnye sudy v Sovetskoi Rossii (1917–1921 gg.)." *Istoriia SSSR* 2 (March–April 1987): 136–47.

Selunskaia, V. M., ed. *Izmeneniia sotsial'noi struktury sovetskogo obshchestva, oktiabr' 1917–1920.* Moscow, Mysl', 1976.

Sovorov, K. I. "Stanovlenie kommunisticheskikh fraktsii v profsoiuzakh (1917–1924 gg.)." *Voprosy istorii KPSS* 12 (1983): 72–81.

Sozinov, E. M. *Kommunisticheskaia partiia v bor'be za perekhod ot rabochego kontrolia k rabochemu upravleniiu i natsionalizatsii promyshlennosti.* Moscow: [n.p.], 1953.

Stishov, M. I. et al., eds. *Iz istorii Velikoi Oktiabr'skoi sotsialisticheskoi revoliutsii: Sbornik statei.* Moscow: Izdatel'stvo Moskovskogo universiteta, 1957.

Strakhov, L. V. "Natsionalizatsiia krupnoi promyshlennosti goroda Moskvy." *Uchenye zapiski Moskovskogo pedagogicheskogo instituta im. V. I. Lenina* 200 (1964): 220–89.

Timofeevski, A. A. et al. *V. I. Lenin i stroitel'stvo partii v pervye gody sovetskoi vlasti.* Moscow: Mysl', 1965.

Trukan, G. A. *Rabochii klass v bor'be za pobedu i prochenie sovetskoi vlasti.* Moscow: Nauka, 1975.

———. *Oktiabr' v tsentral'noi Rossii.* Moscow: Mysl', 1967.

Tsiperovich, G. *Sindikaty i tresty v dorevoliutsionnoi Rossii i v SSSR, izdanie chertvertoe.* Leningrad: Izdatel'stvo tekhnika i proizvodstvo, 1927.

Venediktov, A. V. *Organizatsiia gosudarstvennoi promyshlennosti v SSSR, vol. 1 (1917–1920).* Leningrad: Izdatel'stvo Leningradskogo universiteta, 1957.

Vinogradov, V. A. *Sotsialisticheskoe obobshchestvlenie sredstvy proizvodstva v promyshlennosti SSSR: 1917–1918.* Moscow: Izdatel'stvo Akademii nauk SSSR, 1955.

———. *Voprosy teorii i praktiki sotsialisticheskoi natsionalizatsii promyshlennosti.* Moscow: Nauka, 1964.

———. *Rabochii kontrol' nad proizvodstvom: Teoriia, istoriia, sovremennost'.* Moscow: Nauka, 1983.

Volobuev, P. V. *Proletariat i burzhuaziia Rossii v 1917 g.* Moscow: Mysl', 1964.

Zaozerskaia, E. I. *Rabochaia sila i klassovaia bor'ba na tekstil'nykh manufakturakh Rossii v 20–60 gg. XVIII v.* Moscow: Izdatel'stvo Akademii nauk SSSR, 1960.

Zhuravlev, V. V. *Dekrety sovetskoi vlasti kak istoricheskii istochnik.* Moscow: Nauka, 1979.

———. "Dekrety o zemle, rabochem kontrole i natsionalizatsii bankov." *Istoricheskie zapiski* 100 (1977): 250–84.

Selected Secondary Sources in English

Avrich, Paul. "The Bolshevik Revolution and Workers' Control in Industry." *Slavic Review* 22 (March 1963): 47–63.

Baykov, Alexander. *The Development of the Soviet Economic System.* Cambridge: Cambridge University Press, 1946.

Bettelheim, Charles. *Class Struggle in the USSR, First Period: 1917–1923.* Translated by Brian Pearce. New York: Monthly Review Press, 1976.

Bonnell, Victoria E. *Roots of Rebellion: Workers' Politics and Organizations in St. Petersburg and Moscow, 1900–1914.* Berkeley and Los Angeles: University of California Press, 1983.

Bradley, Joseph. *Muzhik and Muscovite: Urbanization in Late Imperial Russia.* Berkeley: University of California Press, 1985.

Brinton, Maurice. "Factory Committees and the Dictatorship of the Proletariat." *Critique* 4 (Spring 1975): 78–86.

Brovkin, Vladimir. "The Mensheviks' Political Comeback: The Elections to the Provincial City Soviets in Spring 1918." *Russian Review* 42 (January 1983): 1–50.

———. "Politics, Not Economics Was the Key." *Slavic Review* 44 (Summer 1985): 244–50.

Carr, E. H. *The Bolshevik Revolution.* 2 vols. London: Macmillan, 1952.

Chase, William J. *Workers, Society, and the Soviet State: Labor and Life in Moscow, 1918–1929.* Urbana and Chicago: University of Illinois Press, 1987.

———. "Voluntarism, Mobilisation and Coercion: *Subbotniki* 1919–1921." *Soviet Studies* 41 (January 1989): 111–28.

Daniels, Robert. *Red October.* New York: Scribner's, 1967.

Deutscher, Isaac. *Soviet Trade Unions.* London: Oxford University Press, 1950.

Dobb, Maurice. *Soviet Economic Development since 1917.* New York: International Publishers, 1948.

Engelstein, Laura. *Moscow, 1905: Working Class Organization and Political Conflict.* Stanford, Calif.: Stanford University Press, 1982.

Ferro, Marc. *October 1917.* Translated by Norman Stone. London: Routledge and Kegan Paul, 1976.

Fitzpatrick, Sheila. *The Russian Revolution.* New York: Oxford University Press, 1982.

———. "The Bolsheviks' Dilemma: Class, Culture, and Politics in the Early Soviet Years." *Slavic Review* 47 (Winter 1988): 599–613.

Galili, Ziva. *The Menshevik Leaders in the Russian Revolution: Social Realities and Political Strategies.* Princeton, N.J.: Princeton University Press, 1989.

Glickman, Rose L. *Russian Factory Women: Workplace and Society, 1880–1914.* Berkeley: University of California Press, 1984.

Goodey, Chris. "Factory Committees and the Dictatorship of the Proletariat (1918)." *Critique* 3 (Autumn 1974): 27–47.

Haimson, Leopold and Charles Tilly, eds. *Strikes, Wars, and Revolutions in an International Perspective: Strike Waves in the Late Nineteenth and Early Twentieth Centuries.* Cambridge: Cambridge University Press and Paris: Editions de la Maison des Sciences de l'Homme, 1989.

Hasegawa, Tsuyoshi. *The February Revolution: Petrograd, 1917.* Seattle: University of Washington Press, 1981.

Husband, William B. "Workers' Control and Centralization in the Russian Revolution: The Textile Industry of the Central Industrial Region,

1917–1920." *The Carl Beck Papers in Russian and Eastern European Studies*, no. 403 (1985): 1–52.

Husband, William B. "Local Industry in Upheaval: The Ivanovo-Kineshma Textile Strike of 1917." *Slavic Review* 47 (Fall 1988): 448–63.

Johnson, Robert E. *Peasant and Proletarian: The Working Class of Moscow in the Late Nineteenth Century*. New Brunswick, N.J.: Rutgers University Press, 1979.

Kaiser, Daniel, ed. *The Workers' Revolution in Russia, 1917: The View from Below*. Cambridge: Cambridge University Press, 1987.

Kaplan, Frederick. *Bolshevik Ideology and the Ethics of Soviet Labor*. New York: The Philosophical Library, 1968.

Keep, J. L. H. *The Russian Revolution: A Study in Mass Mobilization*. London: Weidenfield and Nicolson, 1976.

Koenker, Diane. "The Evolution of Party Consciousness in 1917: The Case of the Moscow Workers." *Soviet Studies* 30 (1978): 38–62.

———. *Moscow Workers and the 1917 Revolution*. Princeton, N.J.: Princeton University Press, 1981.

Koenker, Diane and William Rosenberg. *Strikes and Revolution in Russia, 1917*. Princeton, N.J.: Princeton University Press, 1989.

McDaniel, Tim. *Autocracy, Capitalism, and Revolution in Russia*. Berkeley and Los Angeles: University of California Press, 1988.

Malle, Silvana. *The Economic Organization of War Communism, 1918–1921*. Cambridge: Cambridge University Press, 1985.

Mandel, David. *The Petrograd Workers and the Fall of the Old Regime: From the February Revolution to the July Days, 1917*. New York: St. Martin's, 1983.

———. *The Petrograd Workers and the Soviet Seizure of Power: From the July Days to July 1918*. New York: St. Martin's, 1984.

Mawdsley, Evan. *The Russian Civil War*. Boston: Allen & Unwin, 1987.

Medvedev, Roy A. *The October Revolution*. Translated by George Saunders. New York: Columbia University Press, 1979.

Owen, Thomas C. *Capitalism and Politics in Russia: A Social History of the Moscow Merchants, 1855–1905*. Cambridge: Cambridge University Press, 1981.

Rabinowitch, Alexander. *The Bolsheviks Come to Power: The Revolution of 1917 in Petrograd*. New York: Norton, 1976.

Raleigh, Donald J. *Revolution on the Volga: 1917 in Saratov*. Ithaca, N.Y.: Cornell University Press, 1986.

Remington, Thomas F. *Building Socialism in Bolshevik Russia: Ideology and Industrial Organization, 1917–1921*. Pittsburgh: Pittsburgh University Press, 1984.

Rieber, Alfred J. *Merchants and Entrepreneurs in Imperial Russia*. Chapel Hill, N.C.: University of North Carolina Press, 1982.

Rigby, T. H. *Lenin's Government: Sovnarkom, 1917–1922*. Cambridge: Cambridge University Press, 1979.

Roberts, Paul Craig. "'War Communism': A Reexamination." *Slavic Review* 29 (June 1970): 238–61.

Rosenberg, William. *Liberals in the Russian Revolution: The Constitutional Democratic Party, 1917–1921*. Princeton, N.J.: Princeton University Press, 1974.

———. "Workers and Workers' Control in the Russian Revolution." *History Workshop* 5 (Spring 1978): 89–97.

———. "The Democratization of Russia's Railroads in 1917." *American Historical Review* 86 (1981): 983–1008.

Rosenberg, William, and Diane Koenker. "Skilled Workers and the Strike Movement in Revolutionary Russia." *Journal of Social History* (Summer 1986): 605–29.

———. "The Limits of Formal Protest: Worker Activism and Social Polarization in Petrograd and Moscow, March to October, 1917." *American Historical Review* 92 (1987): 296–326.

Ruckman, Jo Ann. *The Moscow Business Elite: A Social and Cultural Portrait of Two Generations, 1840–1905*. DeKalb, Ill.: Northern Illinois University Press, 1984.

Sakwa, Richard. *Soviet Communists in Power: A Study of Moscow during the Civil War, 1918–1921*. New York: St. Martin's, 1988.

Service, Robert. *The Bolshevik Party in Revolution: A Study in Organizational Change, 1917–1923*. London: Macmillan, 1979.

Smith, Steven A. *Red Petrograd: Revolution in the Factories, 1917–1918*. Cambridge: Cambridge University Press, 1983.

———. "Craft Consciousness, Class Consciousness: Petrograd 1917." *History Workshop* 11 (Spring 1981): 33–56.

Stites, Richard. *Revolutionary Dreams: Utopian Vision and Experimental Life in the Russian Revolution*. New York: Oxford University Press, 1989.

Suny, Ronald Grigor. *The Baku Commune, 1917–1918: Class and Nationality in the Russian Revolution*. Princeton, N.J.: Princeton University Press, 1972.

———. "Russian Labor and Its Historians in the West: A Report and Discussion of the Berkeley Conference on the Social History of Russian Labor." *International Labor and Working Class History* 22 (Fall 1982): 39–53.

———. "Toward a Social History of the October Revolution." *American Historical Review* 88 (February 1983): 31–52.

Thurston, Robert. *Liberal City, Conservative State: Moscow and Russia's Urban Crisis, 1906–1914*. New York: Oxford University Press, 1987.

Wade, Rex A. *Red Guards and Workers' Militias in the Russian Revolution*. Stanford, Calif.: Stanford University Press, 1984.

Wildman, Allan K. *The Making of a Workers' Revolution*. Chicago: University of Chicago Press, 1967.

———. *The End of the Russian Imperial Army: The Old Army and the Soldiers' Revolt (March–April 1917)*. Princeton, N.J.: Princeton University Press, 1980.

———. *The End of the Russian Imperial Army: The Road to Soviet Power and Peace*. Princeton, N.J.: Princeton University Press, 1987.

Glossary

arshin: a unit of measure equal to approximately 28 inches.

artel' (pl. **arteli**): a cooperative working, eating, and living collective, usually composed of workers from the same village or region.

desiatin: one desiatin equals approximately 2.5 acres.

Duma: the Imperial State Duma, Russia's national representative assembly in 1906–1917.

glavk (pl. **glavki**): a subdivision of *VSNKh*; each sphere of industrial manufacturing was subordinated to either a *glavk* or a directing center.

intelligent (pl. **intelligenty**): in this context, a politically conscious activist opposed to the tsarist regime.

Kadet: member of the Russian Constitutional-Democratic Party, largely a party slightly left of center based in the urban professional classes.

kasha: rolled oats; a staple of the peasant diet.

kombedy: Committees of the Poor; created in June 1918 to alleviate the grain shortage by mobilizing the poorer peasants against the kulaks, who were accused of hoarding grain supplies; largely abandoned in the fall of 1918.

kontrol': authority for accounting and supervision, as opposed to "control" in the sense of hegemony.

kulak: a wealthy peasant who exploits the labor of others; the negative object of Soviet slogans directed to the peasants.

kust (pl. **kusty**): literally "group;" amalgamations of enterprises under a unified administration; each *kust* encompasses all facets of the production cycle of a given product, from raw-material acquisition to distribution.

kustar': cottage, handicraft manufacturing.

muzhik: peasant

oblast': in this context, an administrative division that combines several provinces.

Old Believers: following the schism in the Russian Orthodox Church in the late seventeenth century, this sect of traditionalists rejected church reforms and broke off from the main body of the faithful; shut off from many traditional pursuits, those who remained in Russia entered business in disproportionate numbers.

otdelenie (pl. **otdeleniia**): a section, department, or branch of a parent organization.

otkhodnik (pl. **otkhodniki**): peasants who departed the countryside for seasonal employment elsewhere, often as part of an *artel'*.

pud: a unit of weight equal to approximately 36 pounds.

raion (pl. **raiony**): an administrative subdivision of a province or a major city.

shchi: soup made with cabbage; a staple of the peasant diet.

sluzhashchii (pl. **sluzhashchie**): clerical and white-collar office employees.

sovnarkhoz (pl. **sovnarkhozy**): *oblast'*, provincial, *raion*, and local councils of the national economy technically subordinate to *VSNKh*; regularly competed with soviets for economic authority.

Sovnarkom: Council of People's Commissars; paramount government political organ.

subbotnik (pl. **subbotniki**): voluntary workdays frequently devoted to fuel gathering, plant maintenance, and the like.

tarif: a uniform, minimum pay rate; a term frequently used in textile manufacturing.

tekstil'shchik (pl. **tekstil'shchiki**): wage earner in the textile industry; also adopted by activists who spoke in the name of the textile workers.

uezd (pl. **uezdy**): district; subdivision of a *raion*.

VSNKh (Vesenkha): Supreme Council of the National Economy; created in December 1917 as the agency to assume direction of all facets of supply, distribution, manufacturing, and finance.

zemliachestvo: an informal but important network of village and regional connections that made urban employment opportunities known to the countryside and helped ease the transition to urban life for new arrivals.

Index

Tatishchev, Governor N., 23
Textile industry
  adolescents in, 7, 115, 148
  living and working conditions, 17–
    19, 81–82, 115–17, 147–50
  monopolies, 8, 20–21
  pre-revolutionary history, 7–8, 14–22
  radical reputation of workers in, 19,
    33, 116
  rural ties of workers in, 4, 9, 16, 59,
    117
  women in, 7, 16–17, 21–22, 85–86,
    148–50, 160, 200 *n.* 131
Tekstil'nyi rabochii, 35, 75
Tekstil'shchik
  assessments of organization and
    administration, 60, 108, 141,
    144, 151, 153
  character of unions and, 73–75, 79
  circulation, 119, 154
  one-man management, 141
  personnel shortages, 107
  propaganda and, 117, 128, 144, 157
  reportage of material shortages, 84,
    146–47
  union leadership and, 105, 132
  women presented in, 85, 149
Tomsky, M., 152
Trade Unions. *See also* Union of
    Textile Workers
  All-Russian Council of Trade
    Unions, 53, 60
  First Congress, 53, 66
  Second Congress, 90
  Third Conference, 37, 53
  Union of Metal Workers, 37
  Union of Railroad Workers, 49
Treaty of Brest–Litovsk, 51, 57
Tregubov Factory, 117, 145, 147–48
Trekhgornaia Mill, 18, 37, 61, 70, 113,
    118
Troitskaia Factory, 117, 145
Trotsky, L., 50–52, 124, 126, 138
Trubanov Factory, 145
Tsindel' Factory, 37
Turkestan, 146

Unemployment, 75, 81–82, 116–17,
    122, 150, 165
Union of Textile Workers
  All-Russian Council, 41–44, 57, 75,
    77–78, 103
  Centro-Textile and, 67–74, 78, 91,
    93–97
  conferences, local and regional, 62,
    64, 71, 116, 132, 136, 142, 145,
    152, 154–55
  cultural–educational work, 60, 80–
    81, 117–19, 153–55
  Department of Labor Resources,
    116
  Economic Department, 140
  First Conference, 38
  First Congress, 44, 64–67, 83, 135
  Fourth Congress, 127
  Labor Department, 142
  leadership of, 10, 37, 44, 47, 60, 63,
    75, 91, 105, 107, 132, 193 *n.* 94
  organization, 19, 33, 61, 74–75, 104–
    6, 123–24, 135–36, 151–52
  Organizational–Instructional
    Department, 135, 140, 151–52
  Raion-Textiles and, 78–79
  Second Congress, 96–97, 103–6, 108,
    114, 128
  sections, local and regional
    Bogovsk, 62
    Egor'ev, 113
    Iaroslavl, 142
    Ivanovo–Kineshma regional, 34,
      36–37, 42, 45, 48, 77
    Ivanovo–Voznesensk city, 28, 72–
      73, 115, 155
    Karachev, 151
    Kineshma, 154
    Klintsy, 142
    Moscow city, 37–39, 41, 43–45,
      149, 154
    Moscow *Oblast'*, 63–64, 79, 106
    Moscow provincial, 152
    Naro–Fominsk, 62, 151
    Nefoekhta, 149–50
    Nizhnii--Novgorod, 149